THE ILLUSTRATED GUIDE TO POISONOUS SNAKES

THE ILLUSTRATED GUIDE TO POISONOUS SNAKES

Department of The Army

THE LYONS PRESS

Guilford, Connecticut

An imprint of The Globe Pequot Press

The Lyons Press is an imprint of The Globe Pequot Press.

10 9 8 7 6 5 4 3 2 1

Printed in te United States of America

Library of Congress Cataloging-in-Publication Data

The illustrated guide to poisonous snakes / Department of the Army.
p. cm.
Originally appeared in the U.S. Army survival handbook, 1985 as an appendix called poisonous snakes, with illustrations.
ISBN 1-58574-834-X (pbk. : alk. paper)
1. Poisonous snakes—Identification. 2. Snakebites. I. United States. Dept. of the Army.

QL666.O6I44 2003
597.96'165—dc21

2002043311

TABLE OF CONTENTS

INTRODUCTION

If you fear snakes, it is probably because you are unfamiliar with them or you have the wrong information about them. There is no need for you to fear snakes if you know—

- *Their habits.*
- *How to identify the dangerous kinds.*
- *Precautions to take to prevent snakebite.*
- *What actions to take in case of snakebite.*

For a man wearing shoes and trousers and living in a camp, the danger of being bitten by a poisonous snake is small compared to the hazards of malaria, cholera, dysentery, or other diseases.

Nearly all snakes avoid man if possible. Reportedly, however, a few—the king cobra of Southeast Asia, the bushmaster and tropical rattlesnake of South America, and the mamba of Africa—sometimes aggressively attack man, but even these snakes do so only occasionally. Most snakes get out of the way and are seldom seen.

Ways to Avoid Snakebite

Snakes are widely distributed. They are found in all tropical, subtropical, and most temperate regions. Some species of snakes have specialized glands that contain a toxic venom and long hollow fangs to inject their venom.

Although venomous snakes use their venom to secure food, they also use it for self-defense. Human accidents occur when you don't see or hear the snake, when you step on them, or when you walk too close to them.

Follow these simple rules to reduce the chance of accidental snakebite:

- Don't sleep next to brush, tall grass, large boulders, or trees. They provide hiding places for snakes. Place your sleeping bag in a clearing. Use mosquito netting tucked well under the bag. This netting should provide a good barrier.
- Don't put your hands into dark places, such as rock crevices, heavy brush, or hollow logs, without first investigating.
- Don't step over a fallen tree. Step on the log and look to see if there is a snake resting on the other side.
- Don't walk through heavy brush or tall grass without looking down. Look where you are walking.
- Don't pick up any snake unless you are absolutely positive it is not venomous.
- Don't pick up freshly killed snakes without first severing the head. The nervous system may still be active and a dead snake can deliver a bite.

Snake Groups

Snakes dangerous to man usually fall into two groups: proteroglypha and solenoglypha. Their fangs and their venom best describe these two groups (Figure 1).

Examine the pictures and read the descriptions of snakes in this book.

Group	Fang Type	Venom Type
Proteroglypha	Fixed	Usually dominant neurotoxic
Solenoglypha	Folded	Usually dominant hemotoxic

Figure 1. Snake group characteristics.

Fangs

The proteroglypha have, in front of the upper jaw and preceding the ordinary teeth, permanently erect fangs. These fangs are called fixed fangs.

The solenoglypha have erectile fangs; that is, fangs they can raise to an erect position. These fangs are called folded fangs.

Venom

The fixed-fang snakes (proteroglypha) usually have neurotoxic venoms. These venoms affect the nervous system, making the victim unable to breathe.

The folded-fang snakes (solenoglypha) usually have hemotoxic venoms. These venoms affect the circulatory system, destroying blood cells, damaging skin tissues, and causing internal hemorrhaging.

Remember, however, that most poisonous snakes have both neurotoxic and hemotoxic venom. Usually one type of venom in the snake is dominant and the other is weak.

Poisonous Versus Nonpoisonous Snakes

No single characteristic distinguishes a poisonous snake from a harmless one except the presence of poison fangs and glands. Only in dead specimens can you determine the presence of these fangs and glands without danger.

Descriptions of Poisonous Snakes

There are many different poisonous snakes throughout the world. It is unlikely you will see many except in a zoo. This manual describes only a few poisonous snakes. You should, however, be able to spot a poisonous snake if you—

Learn about the two groups of snakes and the families in which they fall (Figure 2).

Group	Family	Local Effects	Systemic Effects
Solenoglypha *Usually dominant* ***hemotoxic*** *venom affecting the circulatory system*	Viperidae *True vipers with movable front fangs*	Strong pain, swelling, necrosis	Hemorrhaging, internal organ breakdown, destroying of blood cells
	Crotalidae *Pit vipers with movable front fangs*		
	Trimeresurus		
Proteroglypha *Usually dominant* ***neurotoxic*** *venom affecting the nervous system*	Elapidae *Fixed front fangs*		
	Cobra	Various pains, swelling, necrosis	Respiratory collapse
	Krait	No local effects	Respiratory collapse
	Micrurus	Little or no pain; no local symptoms	Respiratory collapse
	Laticaudinae and Hydrophidae *Ocean-living with fixed front fangs*	Pain and local swelling	Respiratory collapse
Note: The venom of the Gaboon viper, the rhinoceros viper, the tropical rattlesnake, and the Mojave rattlesnake is both strongly hemotoxic and strongly neurotoxic.			

Figure 2. Clinical effects of snake bites.

Viperidae

The viperidae or true vipers usually have thick bodies and heads that are much wider than their necks (Figure 3). However, there are many different sizes, markings, and colorations.

This snake group has developed a highly sophisticated means for delivering venom. They have long, hollow fangs that perform like hypodermic needles. They deliver their venom deep into the wound.

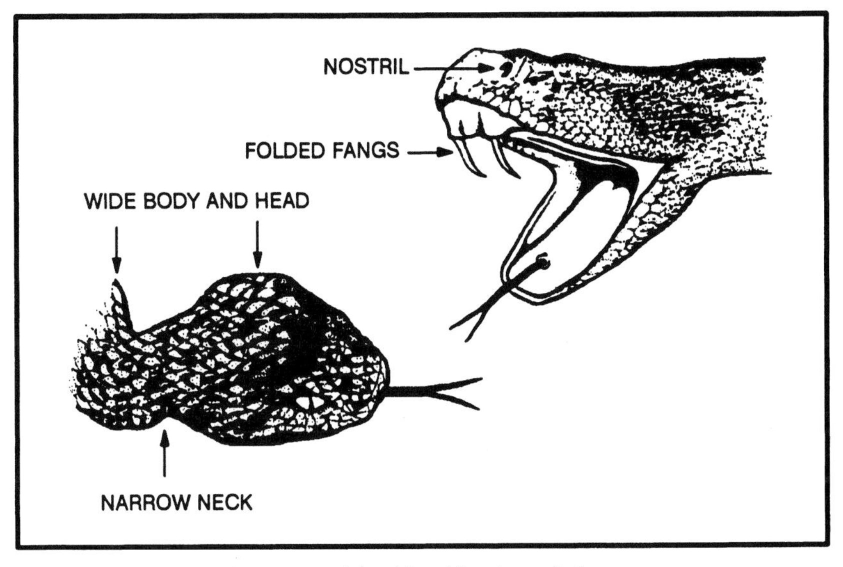

Figure 3. Positive identification of vipers.

The fangs of this group of snakes are movable. These snakes fold their fangs into the roof of their mouths. When they strike, their fangs come forward, stabbing the victim. The snake controls the movement of its fangs; fang movement is not automatic. The venom is usually hemotoxic. There are, however, several species that have large quantities of neurotoxic elements, thus making them even more dangerous. The vipers are responsible for many human fatalities around the world.

Crotalidae

The crotalids, or pit vipers (Figure 4), may be either slender or thick-bodied. Their heads are usually much wider than their necks. These snakes take their name from the deep pit located between the eye and the nostril. They are commonly brown with dark blotches, though some kinds are green.

Rattlesnakes, copperheads, cottonmouths, and several species of dangerous snakes from Central and South America, Asia, China, and India fall into the pit viper group. The pit is a highly sensitive organ capable of picking up the slightest temperature variance. Most pit

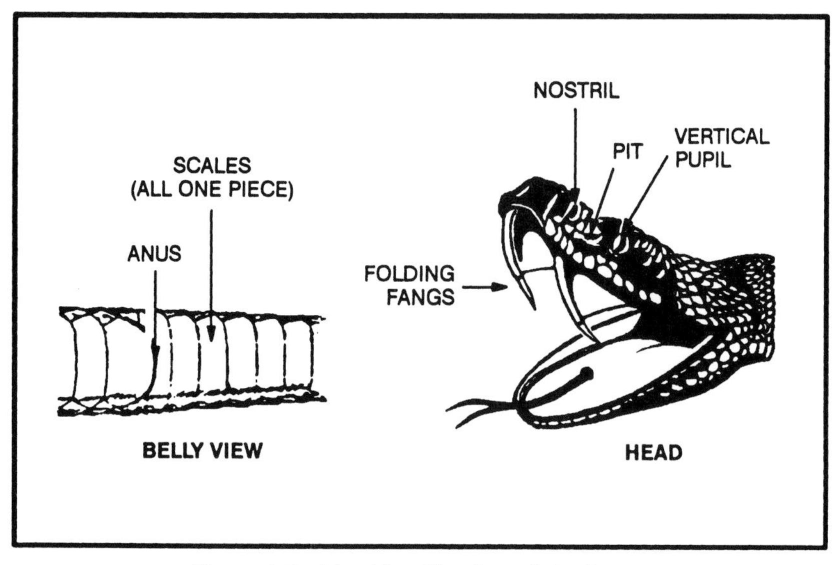

Figure 4. Positive identification of pit vipers.

vipers are nocturnal. They hunt for food at night with the aid of these specialized pits that let them locate prey in total darkness. Rattlesnakes are the only pit vipers that possess a rattle at the tip of the tail.

India has about 12 species of these snakes. You find them in trees or on the ground in all types of terrain. The tree snakes are slender; the ground snakes are heavy-bodied. All are dangerous.

China has a pit viper similar to the cottonmouth found in North America. You find it in the rocky areas of the remote mountains of South China. It reaches a length of 1.4 meters but is not vicious unless irritated. You can also find a small pit viper, about 45 centimeters long, on the plains of eastern China. It is too small to be dangerous to a man wearing shoes.

There are about 27 species of rattlesnakes in the United States and Mexico. They vary in color and may or may not have spots or blotches. Some are small while others, such as the diamondbacks, may grow to 2.5 meters long.

There are five kinds of rattlesnakes in Central and South America, but only the tropical rattlesnake is widely distributed. The rattle on the tip of the tail is sufficient identification for a rattlesnake.

Most will try to escape without a fight when approached, but there is always a chance one will strike at a passerby. They do not always give a warning; they may strike first and rattle afterwards or not at all.

The genus Trimeresurus is a subgroup of the crotalidae. These are Asian pit vipers. These pit vipers are normally tree-loving snakes with a few species living on the ground. They basically have the same characteristics of the crotalidae—slender build and very dangerous. Their bites usually are on the upper extremities—head, neck, and shoulders. Their venom is largely hemotoxic.

Elapidae

Elapids are a group of highly dangerous snakes with powerful neurotoxic venom that affects the nervous system, causing respiratory paralysis. Included in this family are coral snakes, cobras, mambas, and all the Australian venomous snakes. The coral snake is small and has caused human fatalities. The Australian death adder, tiger, taipan, and king brown snakes are among the most venomous in the world, causing many human fatalities.

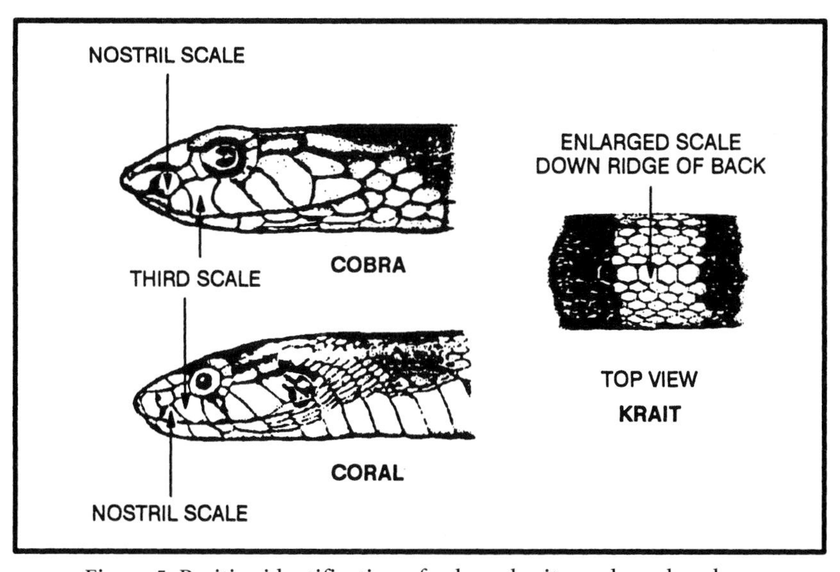

Figure 5. Positive identification of cobras, kraits, and coral snakes.

Only by examining a dead snake can you positively determine if it is a cobra or a near relative (Figure 5). On cobras, kraits, and coral snakes, the third scale on the upper lip touches both the nostril scale and the eye. The krait also has a row of enlarged scales down its ridged back.

You can find the cobras of Africa and the Near East in almost any habitat. One kind may live in or near water, another in trees. Some are aggressive and savage. The distance a cobra can strike in a forward direction is equal to the distance its head is raised above the ground. Some cobras, however, can spit venom a distance of 3 to 3.5 meters. This venom is harmless unless it gets into your eyes; then it may cause blindness if not washed out immediately. Poking around in holes and rock piles is dangerous because of the chance of encountering a spitting cobra.

Laticaudinae and Hydrophidae

A subfamily of elapidae, these snakes are specialized in that they found a better environment in the oceans. Why they are in the oceans is not clear to science.

Sea snakes differ in appearance from other snakes in that they have an oarlike tail to aid in swimming. Some species of sea snakes have venom several times more toxic than the cobra's. Because of their marine environment, sea snakes seldom come in contact with humans. The exceptions are fishermen who capture these dangerous snakes in fish nets and scuba divers who swim in waters where sea snakes are found.

There are many species of sea snakes. They vary greatly in color and shape. Their scales distinguish them from eels that have no scales.

Sea snakes occur in salt water along the coasts throughout the Pacific. There are also sea snakes on the east coast of Africa and in the Persian Gulf. There are no sea snakes in the Atlantic Ocean.

There is no need to fear sea snakes. They have not been known to attack a man swimming. Fishermen occasionally get bit by a sea snake caught in a net. The bite is dangerous.

Colubridae

Colubrids are the largest group of snakes worldwide. In this family there are species that are rear-fanged; however, most are completely harmless to man.

They have a venom-producing gland and enlarged, grooved rear fangs that allow venom to flow into the wound. The inefficient venom apparatus and the specialized venom is effective on cold-blooded animals (such as frogs and lizards) but not considered a threat to human life. The boomslang and the twig snake of Africa have, however, caused human deaths.

Viperidae	Common adder
	Long-nosed adder
	Gaboon viper
	Levant viper
	Horned desert viper
	McMahon's viper
	Mole viper
	Palestinian viper
	Puff adder
	Rhinoceros viper
	Russell's viper
	Sand viper
	Saw-scaled viper
	Ursini's viper
Elapidae	Australian copperhead
	Common cobra
	Coral snake
	Death adder
	Egyptian cobra
	Green mamba
	King cobra
	Krait
	Taipan
	Tiger snake
Crotalidae	American copperhead
	Boomslang
	Bush viper
	Bushmaster
	Cottonmouth
	Eastern diamondback rattlesnake
	Eyelash pit viper
	Fer-de-lance
	Green tree pit viper
	Habu pit viper
	Jumping viper
	Malayan pit viper
	Mojave rattlesnake
	Pallas' viper
	Tropical rattlesnake
	Wagler's pit viper
	Western diamondback rattlesnake
Hydrophidae	Banded sea snake
	Yellow-bellied sea snake

POISONOUS SNAKES OF THE AMERICAS

American copperhead
Agkistrodon contortrix

Description: Chestnut color dominates overall, with darker crossbands of rich browns that become narrower on top and widen at the bottom. The top of the head is a coppery color.

Characteristics: Very common over much of its range, with a natural camouflage ability to blend in the environment. Copperheads are rather quiet and inoffensive in disposition but will defend themselves vigorously. Bites occur when the snakes are stepped on or when a victim is lying next to one. A copperhead lying on a bed of dead leaves becomes invisible. Its venom is hemotoxic.

Habitat: Found in wooded and rocky areas and mountainous regions.

Length: Average 60 centimeters, maximum 120 centimeters.

Distribution: Eastern Gulf States, Texas, Arkansas, Maryland, north Florida, Illinois, Oklahoma, Kansas, Ohio, New York, Alabama, Tennessee, and Massachusetts.

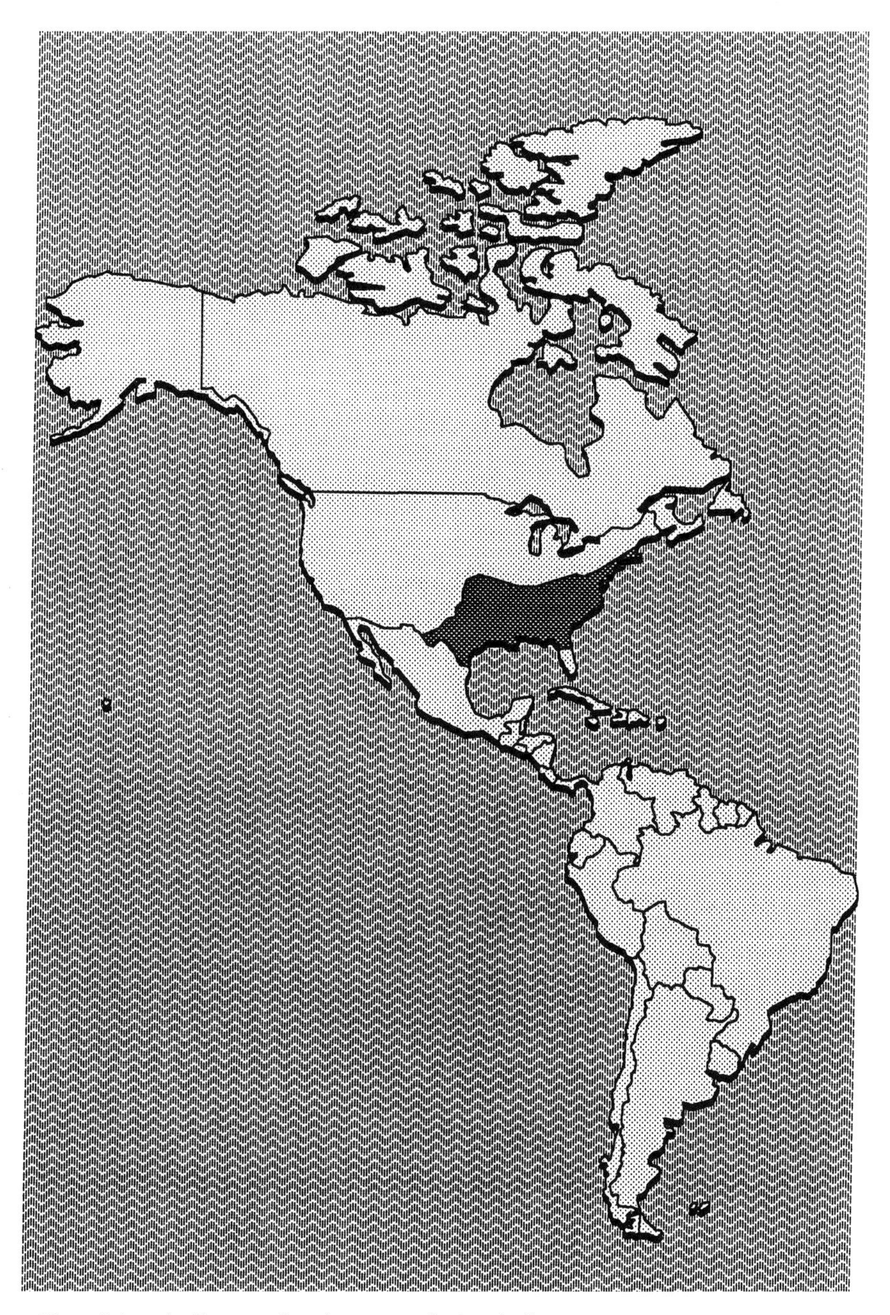

Note: Maps indicate only where population is densest.

Bushmaster

Lachesis mutus

Description: The body hue is rather pale brown or pinkish, with a series of large bold dark brown or black blotches extending along the body. Its scales are extremely rough.

Characteristics: The world's largest pit viper has a bad reputation. This huge venomous snake is not common anywhere in its range. It lives in remote and isolated habitats and is largely nocturnal in its feeding habits; it seldom bites anyone, so few bites are recorded. A bite from one would indeed be very serious and fatal if medical aid was not immediately available. Usually, the bites occur in remote, dense jungles, many kilometers and several hours or even days away from medical help. Bushmaster fangs are long. In large bushmasters, they can measure 3.8 centimeters. Its venom is a powerful hemotoxin.

Habitat: Found chiefly in tropical forests in their range.

Length: Average 2.1 meters, maximum 3.7 meters.

Distribution: Nicaragua, Costa Rica, Panama, Trinidad, and Brazil.

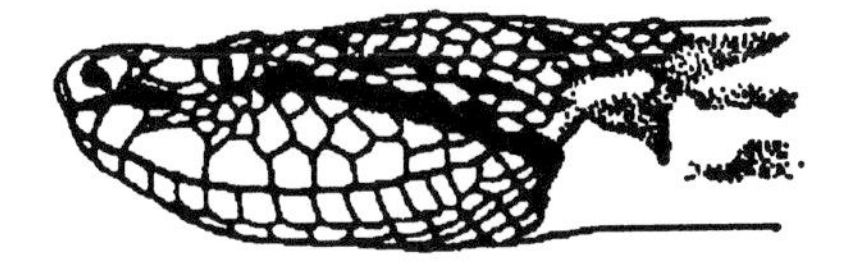

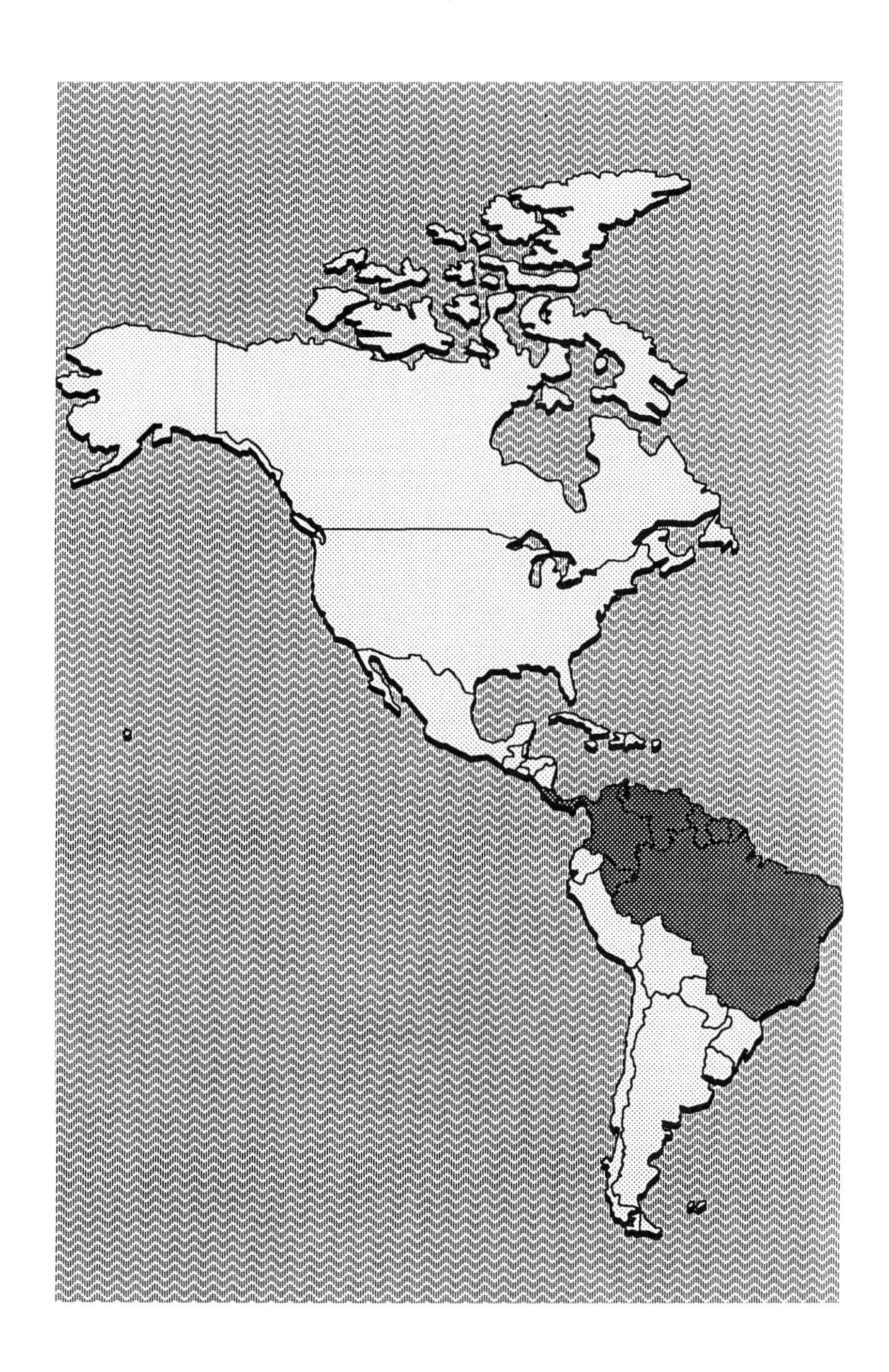

Coral snake

Micrurus fulvius

Description: Beautifully marked with bright blacks, reds, and yellows. To identify the species, remember that when red touches yellow it is a coral snake.

Characteristics: Common over range, but secretive in its habits, therefore seldom seen. It has short fangs that are fixed in an erect position. It often chews to release its venom into a wound. Its venom is very powerful. The venom is neurotoxic, causing respiratory paralysis in the victim, who succumbs to suffocation.

Habitat: Found in a variety of habitats including wooded areas, swamps, palmetto and scrub areas. Coral snakes often venture into residential locations.

Length: Average 60 centimeters, maximum 115 centimeters.

Distribution: Southeast North Carolina, Gulf States, west central Mississippi, Florida, Florida Keys and west to Texas. Another genus of coral snake is found in Arizona. Coral snakes are also found throughout Central and most of South America.

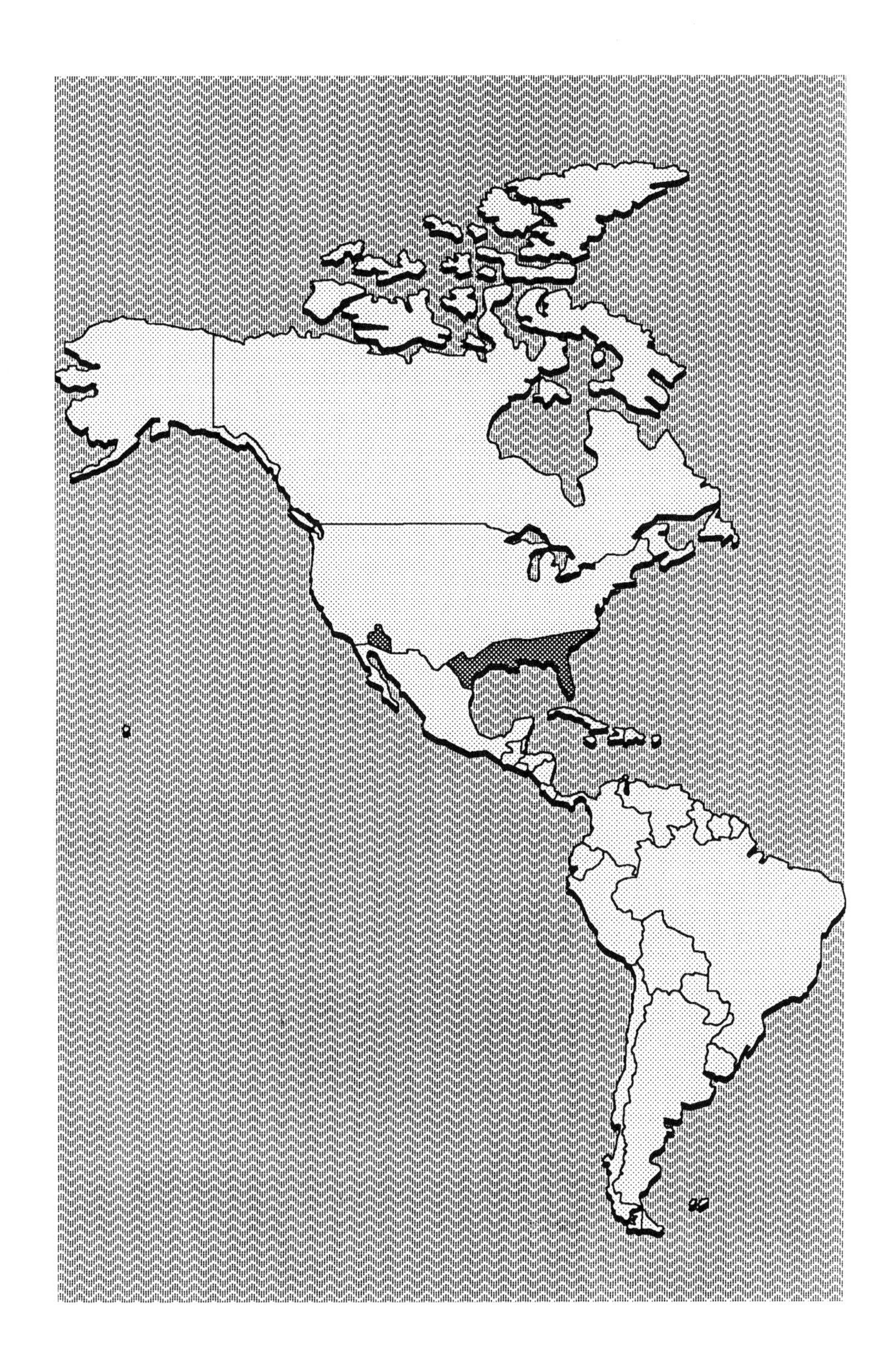

Cottonmouth

Agkistrodon piscivorus

Description: Colors are variable. Adults are uniformly olive brown or black. The young and subadults are strongly crossbanded with dark brown.

Characteristics: These dangerous semiaquatic snakes closely resemble harmless water snakes that have the same habitat. Therefore, it is best to leave all water snakes alone. Cottonmouths often stand their ground. An aroused cottonmouth will draw its head close to its body and open its mouth showing its white interior. Cottonmouth venom is hemotoxic and potent. Bites are prone to gangrene.

Habitat: Found in swamps, lakes, rivers, and ditches.

Length: Average 90 centimeters, maximum 1.8 meters.

Distribution: Southeast Virginia, west central Alabama, south Georgia, Illinois, east central Kentucky, south central Oklahoma, Texas, North and South Carolina, Florida, and the Florida Keys.

Eastern diamondback rattlesnake
Crotalus adamanteus

Description: Diamonds are dark brown or black, outlined by a row of cream or yellowish scales. Ground color is olive to brown.

Characteristics: The largest venomous snake in the United States. Large individual snakes can have fangs that measure 2.5 centimeters in a straight line. This species has a sullen disposition, ready to defend itself when threatened. Its venom is potent and hemotoxic, causing great pain and damage to tissue.

Habitat: Found in palmettos and scrubs, swamps, pine woods, and flatwoods. It has been observed swimming many miles out in the Gulf of Mexico, reaching some of the islands off the Florida coast.

Length: Average 1.4 meters, maximum 2.4 meters.

Distribution: Coastal areas of North Carolina, South Carolina, Louisiana, Florida, and the Florida Keys.

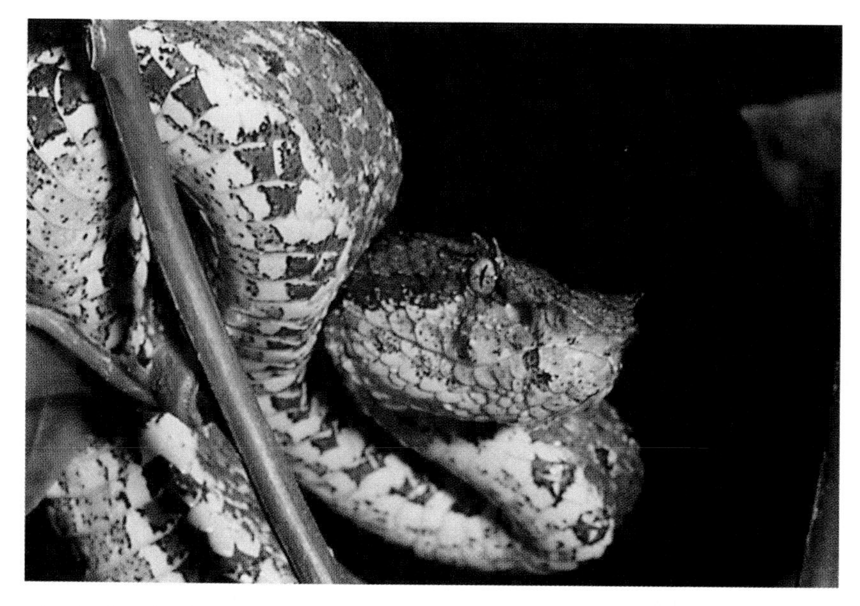

Eyelash pit viper

Bothrops schlegeli

Description: Identified by several spiny scales over each eye. Color is highly variable, from bright yellow over its entire body to reddish-yellow spots throughout the body.

Characteristics: Arboreal snake that seldom comes to the ground. It feels more secure in low-hanging trees where it looks for tree frogs and birds. It is a dangerous species because most of its bites occur on the upper extremities. It has an irritable disposition. It will strike with little provocation. Its venom is hemotoxic, causing severe tissue damage. Deaths have occurred from the bites of these snakes.

Habitat: Tree-loving species found in rain forests; common on plantations and in palm trees.

Length: Average 45 centimeters, maximum 75 centimeters.

Distribution: Southern Mexico, throughout Central America, Columbia, Ecuador, and Venezuela.

Fer-de-lance

Bothrops atrox

There are several closely related species in this group. All are very dangerous to man.

Description: Variable coloration, from gray to olive, brown, or reddish, with dark triangles edged with light scales. Triangles are narrow at the top and wide at the bottom.

Characteristics: This highly dangerous snake is responsible for a high mortality rate. It has an irritable disposition, ready to strike with little provocation. The female fer-de-lance is highly prolific, producing up to 60 young born with a dangerous bite. The venom of this species is hemotoxic, painful, and hemorrhagic (causing profuse internal bleeding). The venom causes massive tissue destruction.

Habitat: Found on cultivated land and farms, often entering houses in search of rodents.

Length: Average 1.4 meters, maximum 2.4 meters.

Distribution: Southern Mexico, throughout Central and South America.

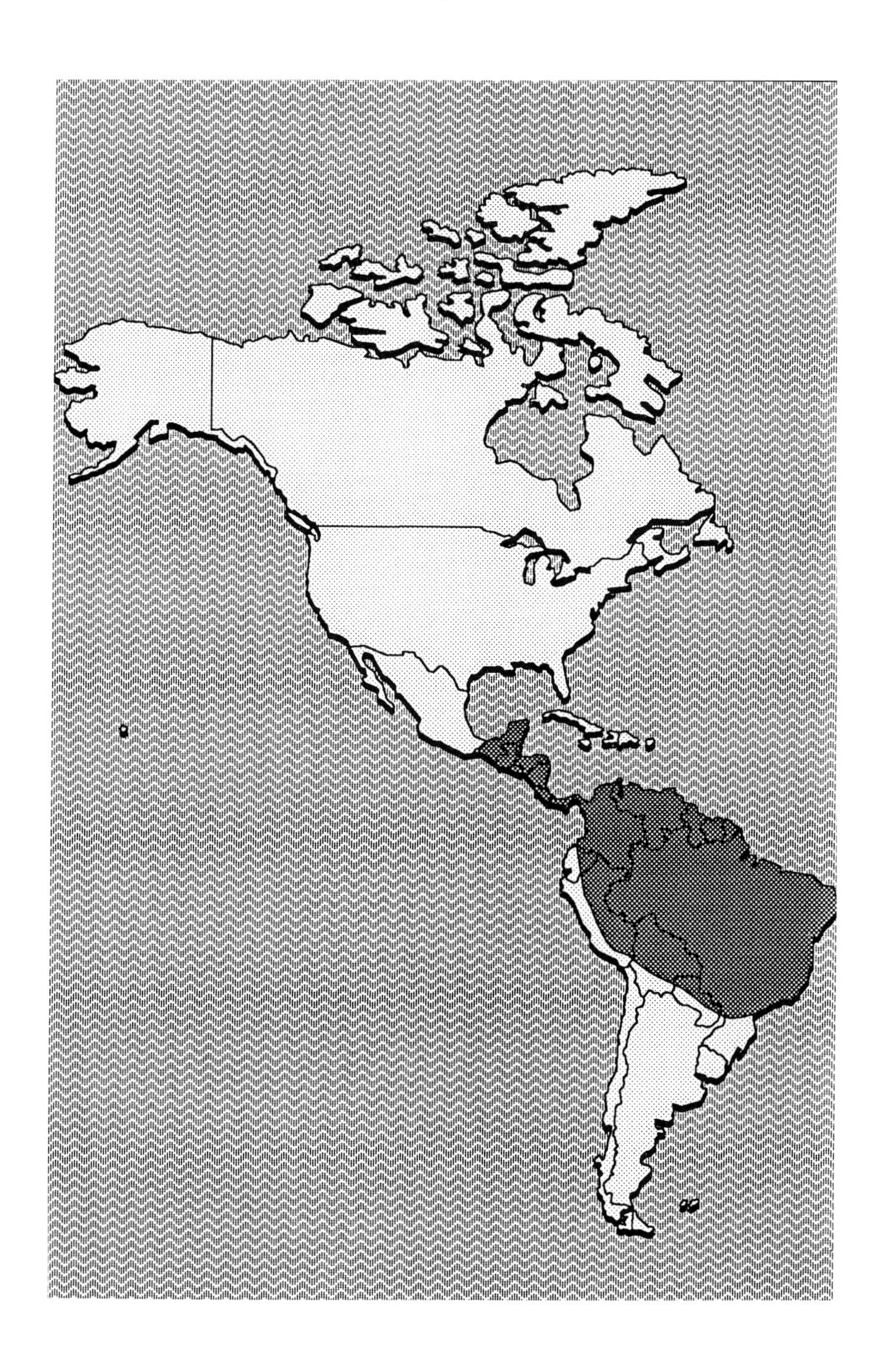

Jumping viper

Bothrops nummifer

Description: It has a stocky body. Its ground color varies from brown to gray and it has dark brown or black dorsal blotches. It has no pattern on its head.

Characteristics: It is chiefly a nocturnal snake. It comes out in the early evening hours to feed on lizards, rodents, and frogs. As the name implies, this species can strike with force as it actually leaves the ground. Its venom is hemotoxic. Humans have died from the bites inflicted by large jumping vipers. They often hide under fallen logs and piles of leaves and are difficult to see.

Habitat: Found in rain forests, on plantations, and on wooded hillsides.

Length: Average 60 centimeters, maximum 120 centimeters.

Distribution: Southern Mexico, Honduras, Guatemala, Costa Rica, Panama, and El Salvador.

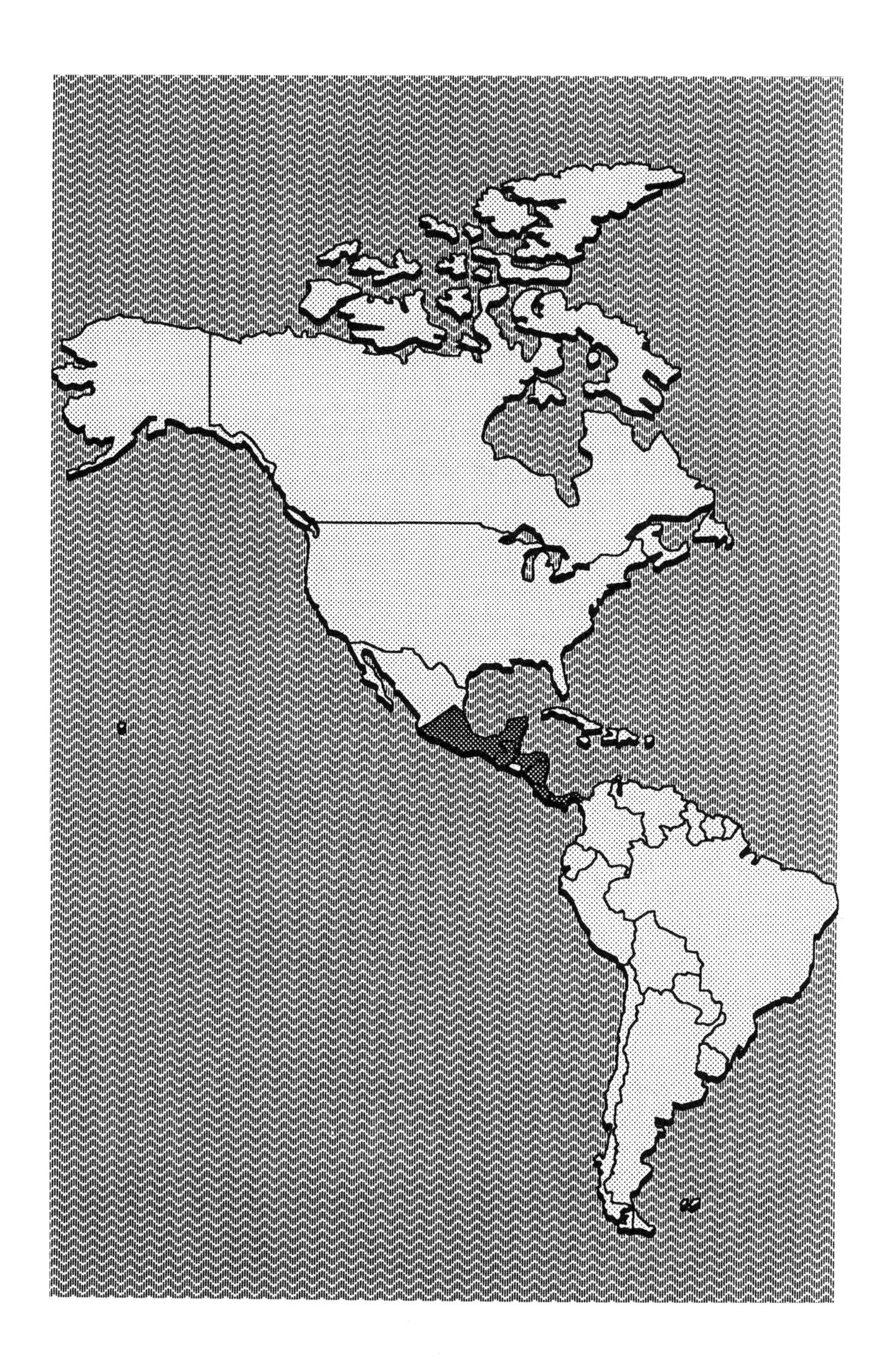

Mojave rattlesnake

Crotalus scutulatus

Description: This snake's entire body is a pallid or sandy color with darker diamond-shaped markings bordered by lighter-colored scales and black bands around the tail.

Characteristics: Although this rattlesnake is of moderate size, its bite is very serious. Its venom has quantities of neurotoxic elements that affect the central nervous system. Deaths have resulted from this snake's bite.

Habitat: Found in arid regions, deserts, and rocky hillsides from sea level to 2400-meter elevations.

Length: Average 75 centimeters, maximum 1.2 meters.

Distribution: Mojave Desert in California, Nevada, southwest Arizona, and Texas into Mexico.

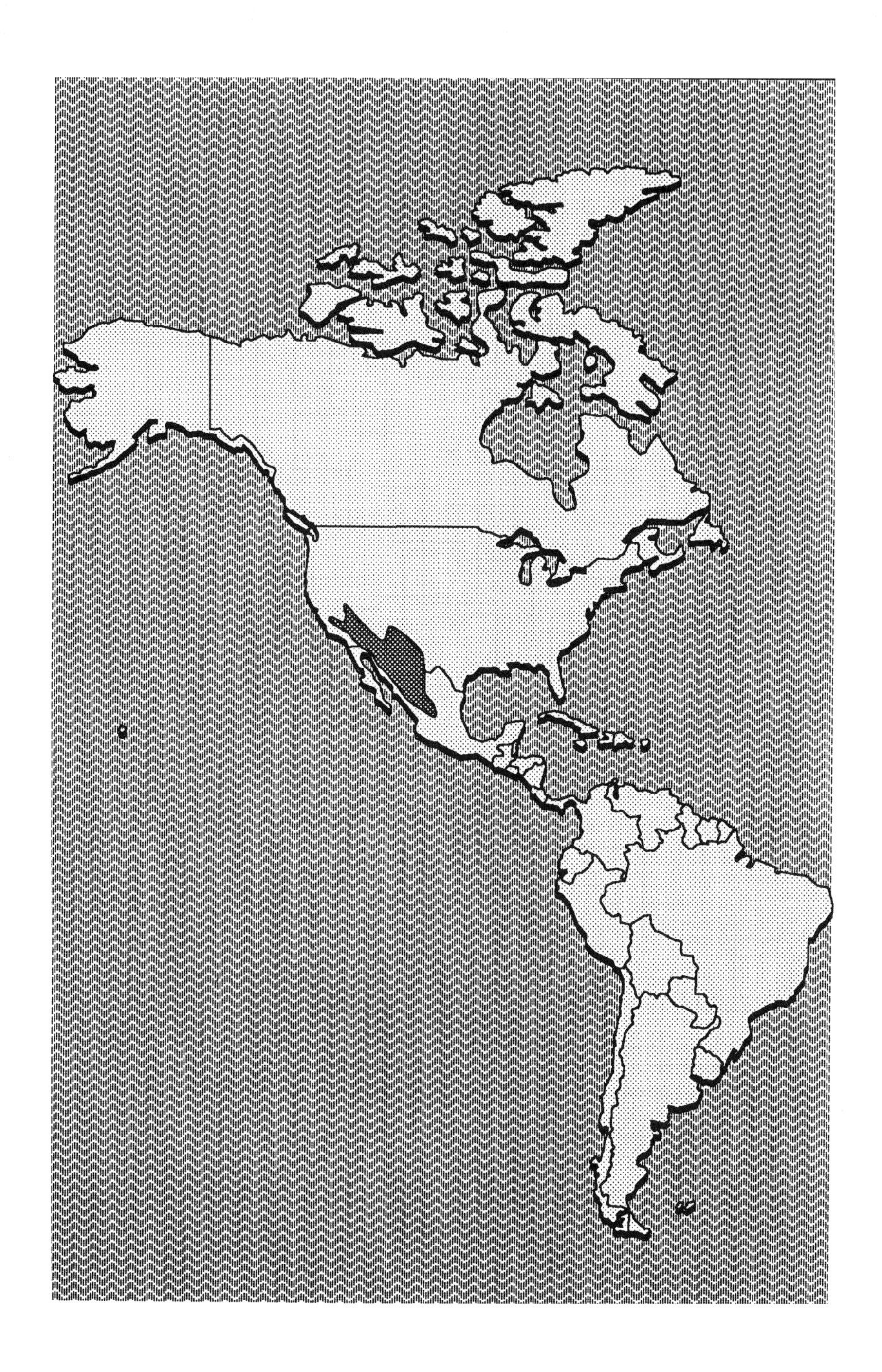

Tropical rattlesnake

Crotalus terrificus

Description: Coloration is light to dark brown with a series of darker rhombs or diamonds bordered by a buff color.

Characteristic: Extremely dangerous with an irritable disposition, ready to strike with little or no warning (use of its rattle). This species has a highly toxic venom containing neurotoxic and hemotoxic components that paralyze the central nervous system and cause great damage to tissue.

Habitat: Found in sandy places, plantations, and dry hillsides.

Length: Average 1.4 meters, minimum 2.1 meters.

Distribution: Southern Mexico, Central America, and Brazil to Argentina.

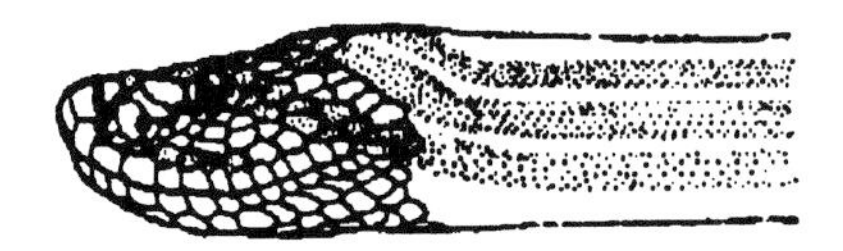

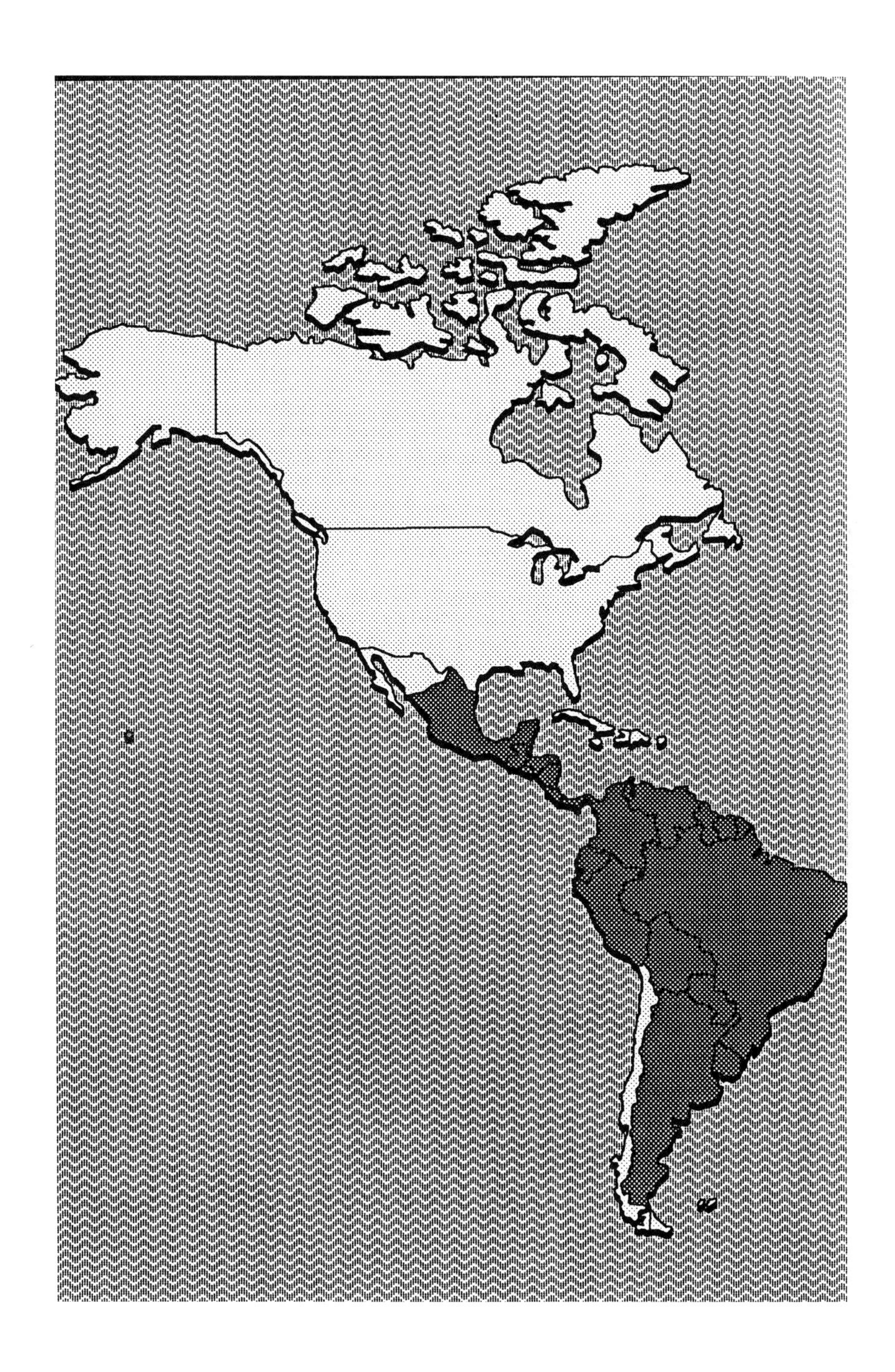

Western diamondback rattlesnake

Crotalus atrox

Description: The body is a light buff color with darker brown diamond-shaped markings. The tail has heavy black and white bands.

Characteristics: This bold rattlesnake holds its ground. When coiled and rattling, it is ready to defend itself. It injects a large amount of venom when it bites, making it one of the most dangerous snakes. Its venom is hemotoxic, causing considerable pain and tissue damage.

Habitat: It is a very common snake over its range. It is found in grasslands, deserts, woodlands, and canyons.

Length: Average 1.5 meters, maximum 2 meters.

Distribution: Southeast California, Oklahoma, Texas, New Mexico, and Arizona.

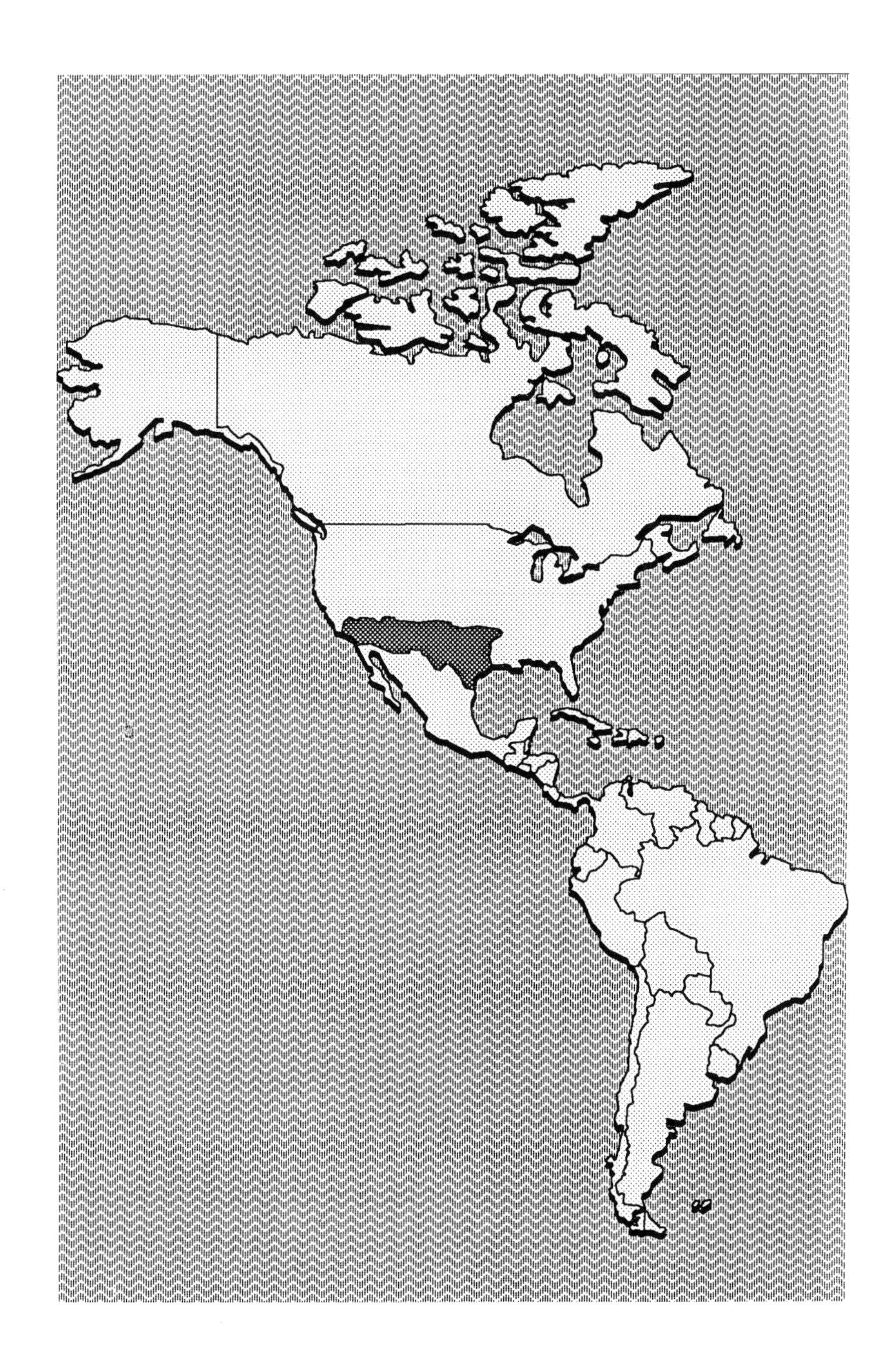

POISONOUS SNAKES OF EUROPE

Common adder

Vipera berus

Description: Its color is variable. Some adult specimens are completely black while others have a dark zigzag pattern running along the back.

Characteristics: The common adder is a small true viper that has a short temper and often strikes without hesitation. Its venom is hemotoxic, destroying blood cells and causing tissue damage. Most injuries occur to campers, hikers, and field workers.

Habitat: Common adders are found in a variety of habitats, from grassy fields to rocky slopes, and on farms and cultivated lands.

Length: Average 45 centimeters, maximum 60 centimeters.

Distribution: Very common throughout most of Europe.

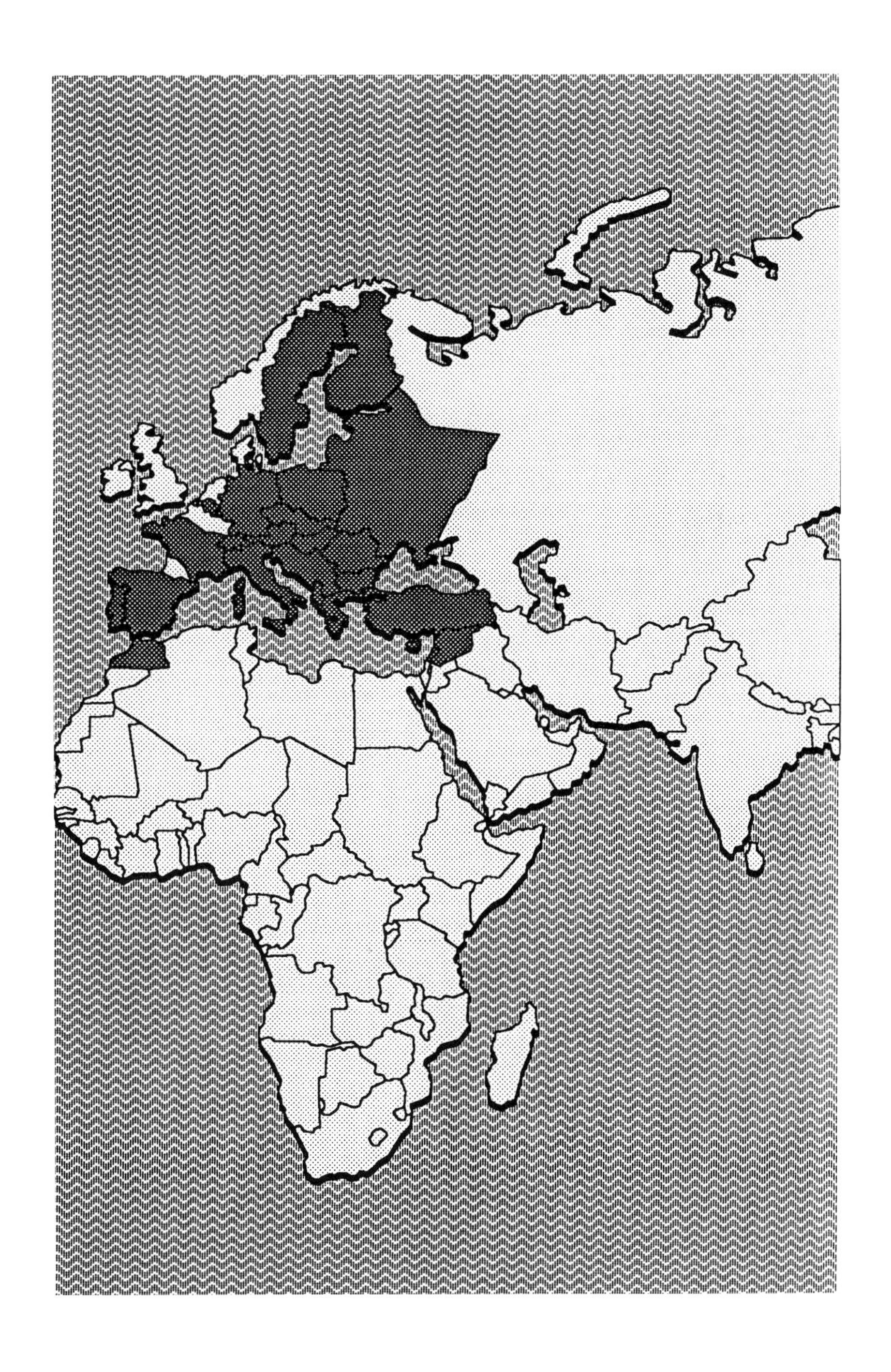

Long-nosed adder

Vipera ammodytes

Description: Coloration is gray, brown, or reddish with a dark brown or black zigzag pattern running the length of its back. A dark stripe is usually found behind each eye.

Characteristics: A small snake commonly found in much of its range. The term "long-nosed" comes from the projection of tiny scales located on the tip of its nose. This viper is responsible for many bites. Deaths have been recorded. Its venom is hemotoxic, causing severe pain and massive tissue damage. The rate of survival is good with medical aid.

Habitat: Open fields, cultivated lands, farms, and rocky slopes.

Length: Average 45 centimeters, maximum 90 centimeters.

Distribution: Italy, Yugoslavia, northern Albania, and Romania.

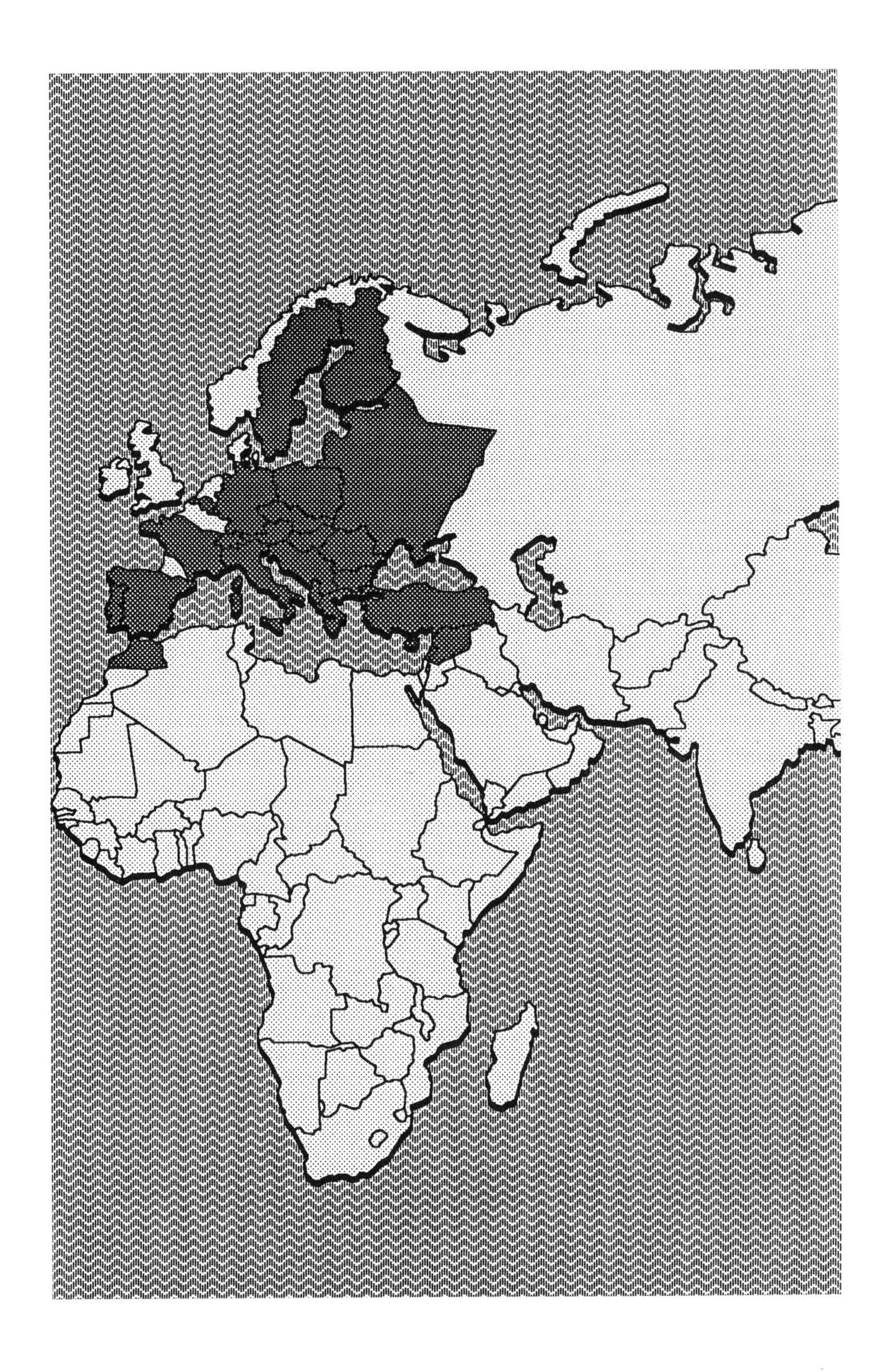

JOHN H. TASHJIAN/BERND VON SCHROEDER

Pallas' viper
Agkistrodon halys

Description: Coloration is gray, tan, or yellow, with markings similar to those of the American copperhead.

Characteristics: This snake is timid and rarely strikes. Its venom is hemotoxic but rarely fatal.

Habitat: Found in open fields, hillsides, and farming regions.

Length: Average 45 centimeters, maximum 90 centimeters.

Distribution: Throughout southeastern Europe.

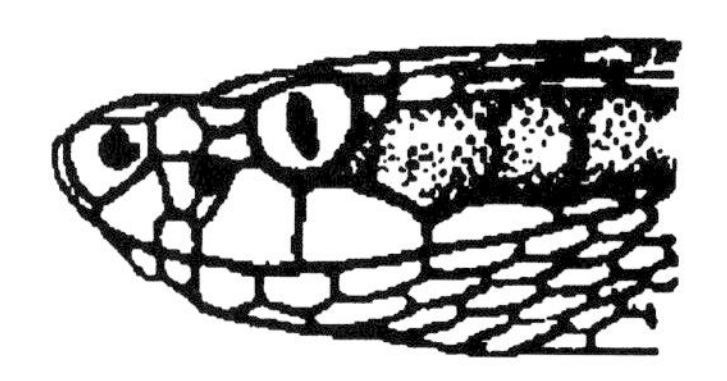

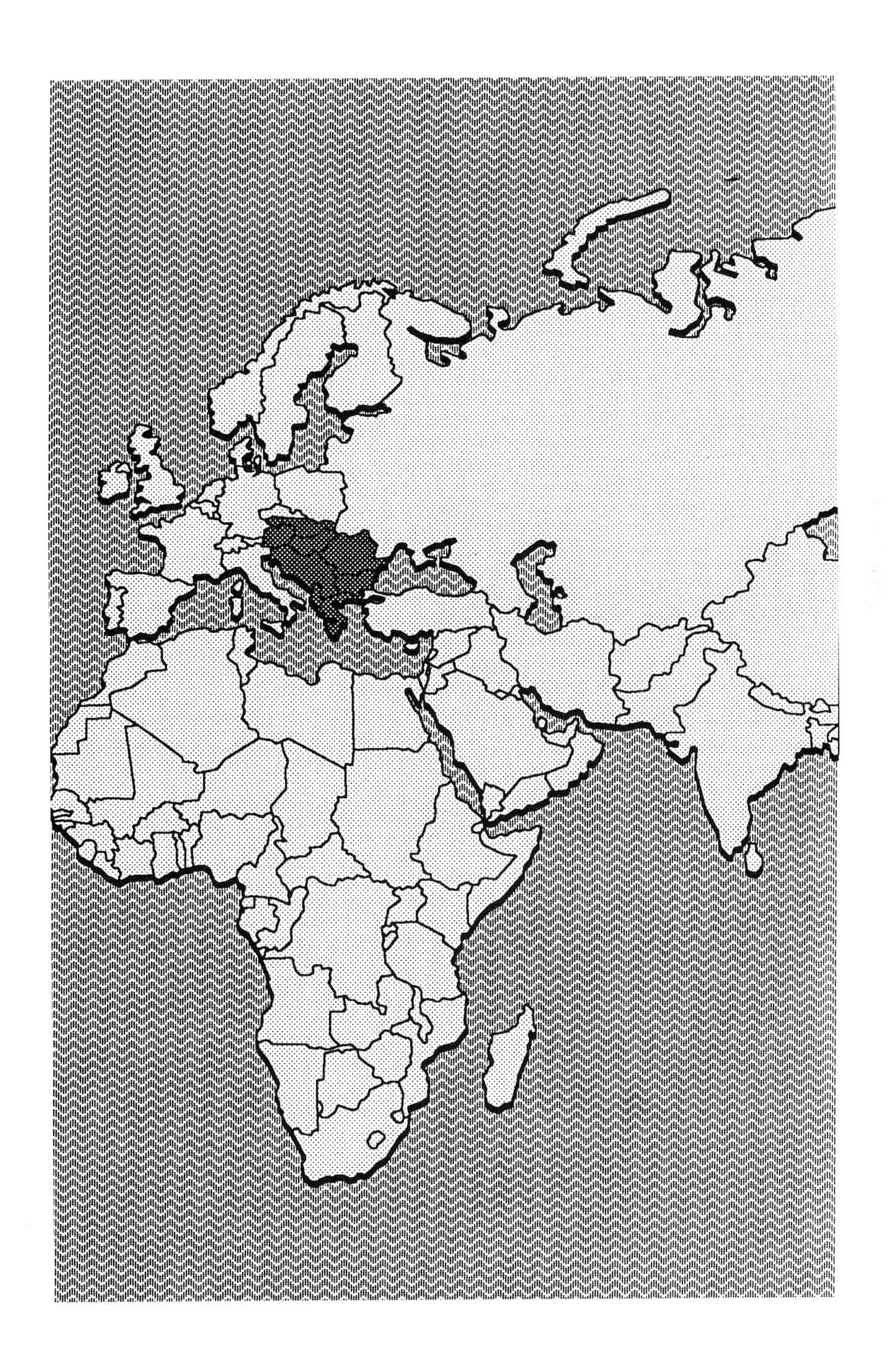

JOHN H. TASHJIAN/BÖRJE FLARDH

Ursini's viper

Vipera ursinii

Description: The common adder, long-nosed adder, and Ursini's viper basically have the same coloration and dorsal zigzag pattern. The exception among these adders is that the common adder and Ursini's viper lack the projection of tiny scales on the tip of the nose.

Characteristics: These little vipers have an irritable disposition. They will readily strike when approached. Their venom is hemotoxic. Although rare, deaths from the bites of these vipers have been recorded.

Habitat: Meadows, farmlands, rocky hillsides, and open, grassy fields.

Length: Average 45 centimeters, maximum 90 centimeters.

Distribution: Most of Europe, Greece, Germany, Yugoslavia, France, Italy, Hungary, Romania, Bulgaria, and Albania.

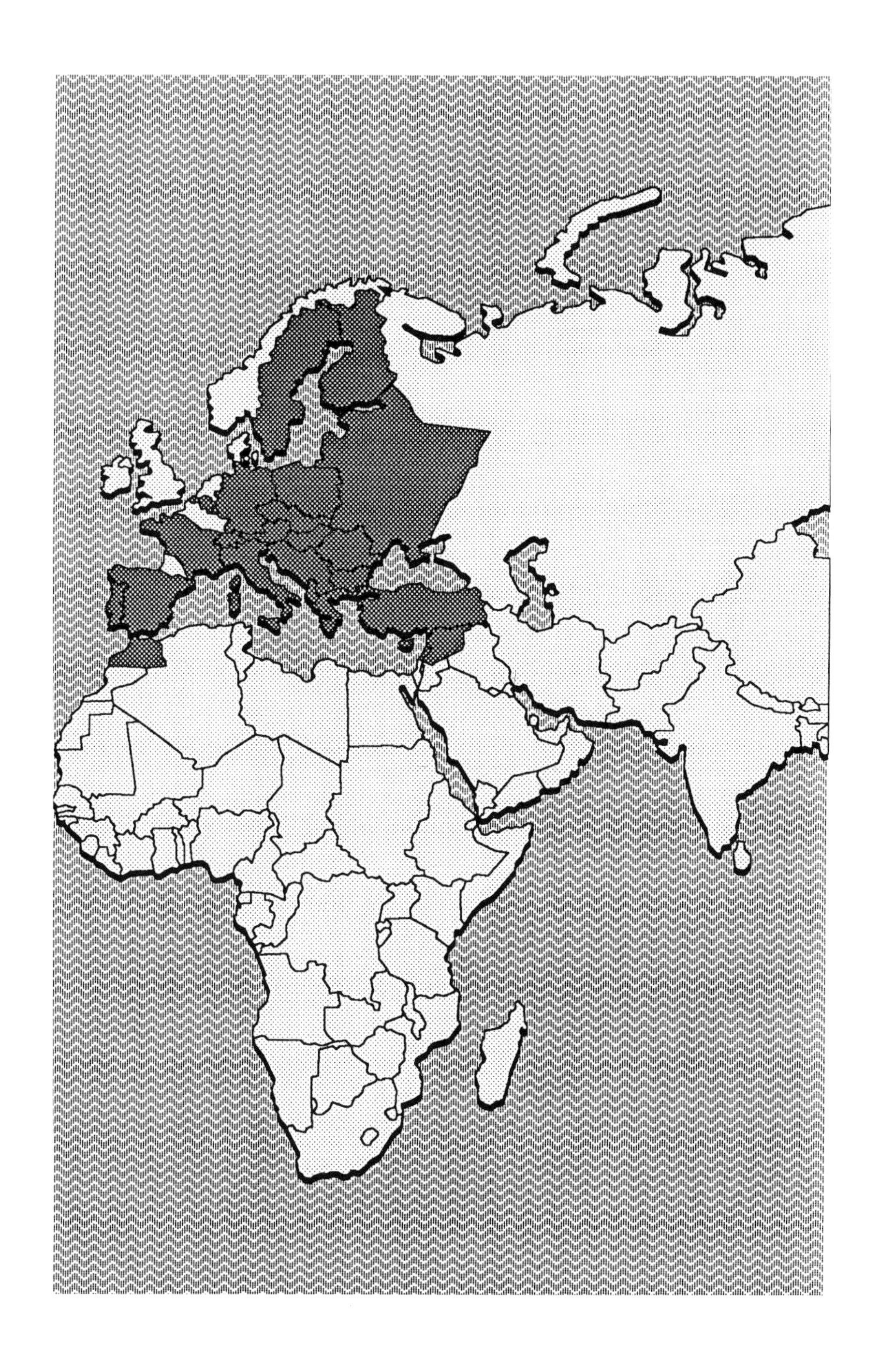

POISONOUS SNAKES OF AFRICA AND ASIA

JOHN H. TASHJIAN/CALIFORNIA ACADEMY OF SCIENCES

Boomslang

Dispholidus typus

Description: Coloration varies but is generally green or brown, which makes it very hard to see in its habitat.

Characteristics: Will strike if molested. Its venom is hemotoxic; even small amounts cause severe hemorrhaging, making it dangerous to man.

Habitat: Found in forested areas. It will spend most of its time in trees or looking for chameleons and other prey in bushes.

Length: Generally less than 60 centimeters.

Distribution: Found throughout sub-Saharan Africa.

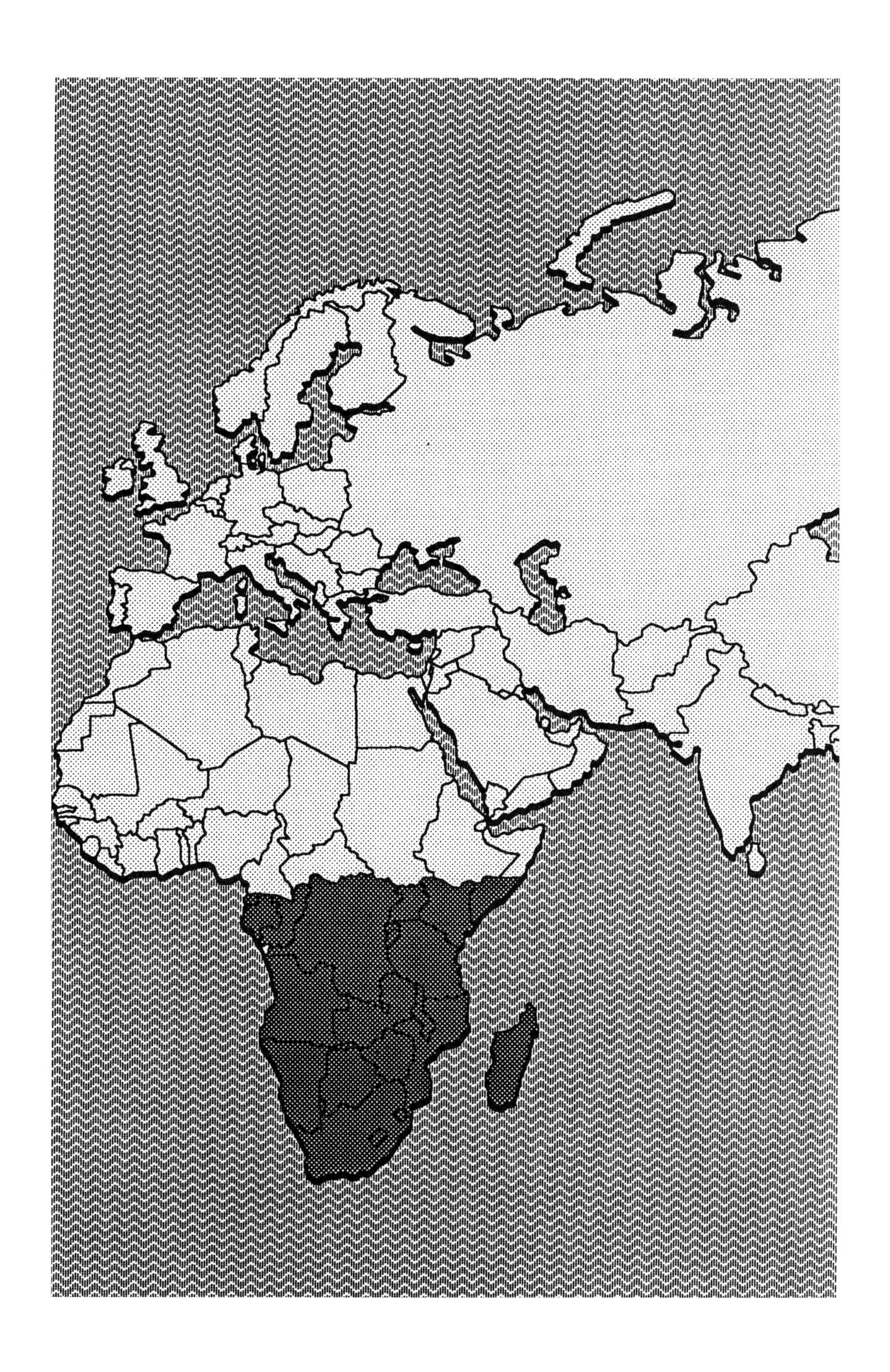

Bush viper
Atheris squamiger

Description: Often called leaf viper, its color varies from ground colors of pale green to olive, brown, or rusty brown. It uses its prehensile tail to secure itself to branches.

Characteristics: An arboreal species that often comes down to the ground to feed on small rodents. It is not aggressive, but it will defend itself when molested or touched. Its venom is hemotoxic; healthy adults rarely die from its bite.

Habitat: Found in rain forests and woodlands bordering swamps and forests. Often found in trees, low-hanging branches, or brush.

Length: Average 45 centimeters, maximum 75 centimeters.

Distribution: Most of Africa, Angola, Cameroon, Uganda, Kenya, and Zaire.

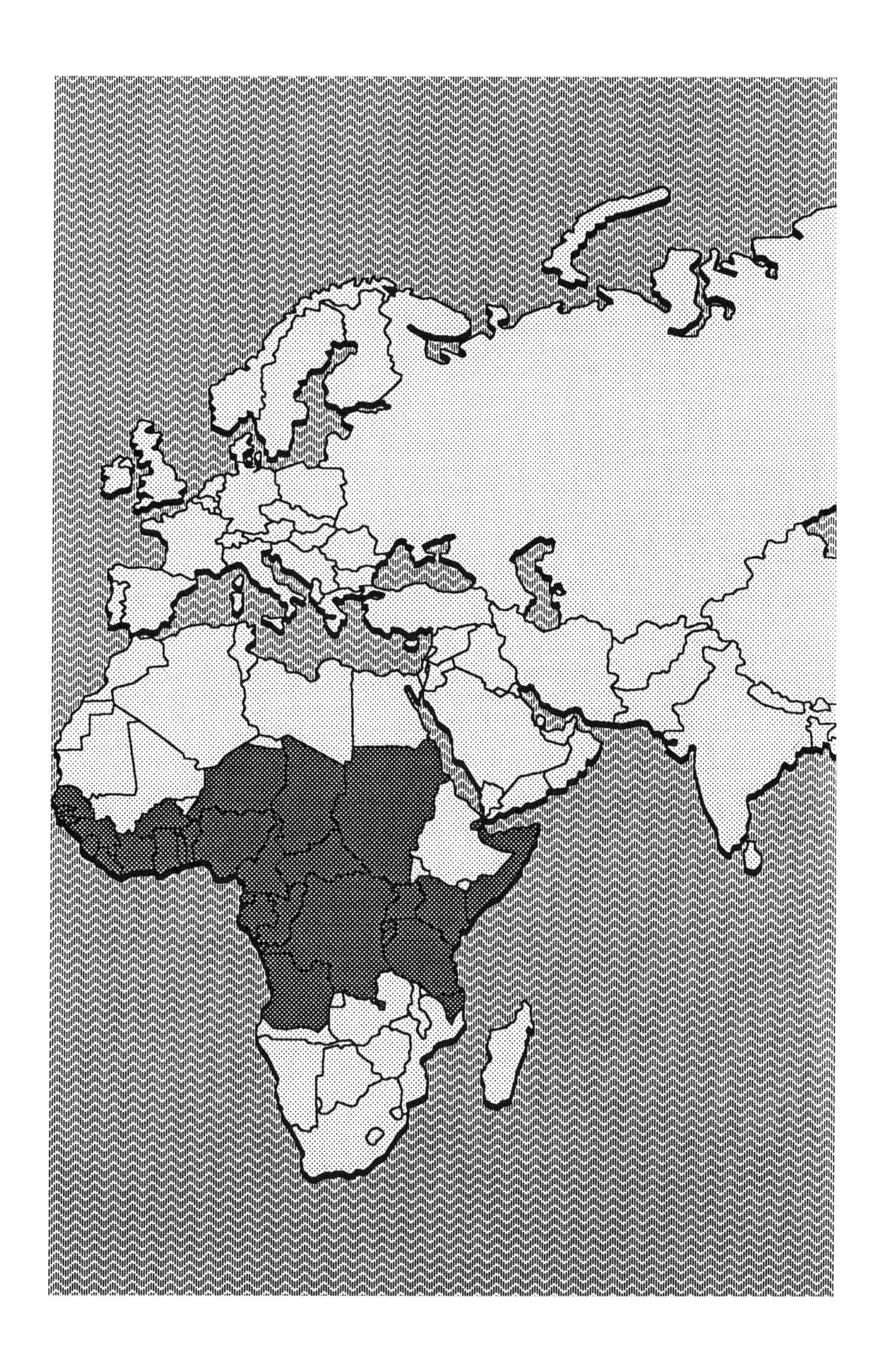

Common cobra

Naja naja

Description: Also known as the Asiatic cobra. Usually slate gray to brown overall. The back of the hood may or may not have a pattern.

Characteristics: A very common species responsible for many deaths each year. When aroused or threatened, the cobra will lift its head off the ground and spread its hood, making it more menacing. Its venom is highly neuro-toxic, causing respiratory paralysis with some tissue damage. The cobra would rather retreat if possible, but if escape is shut off, it will be a dangerous creature to deal with.

Habitat: Found in any habitat: cultivated farms, swamps, open fields, and human dwellings where it searches for rodents.

Length: Average 1.2 meters, maximum 2.1 meters.

Distribution: All of Asia.

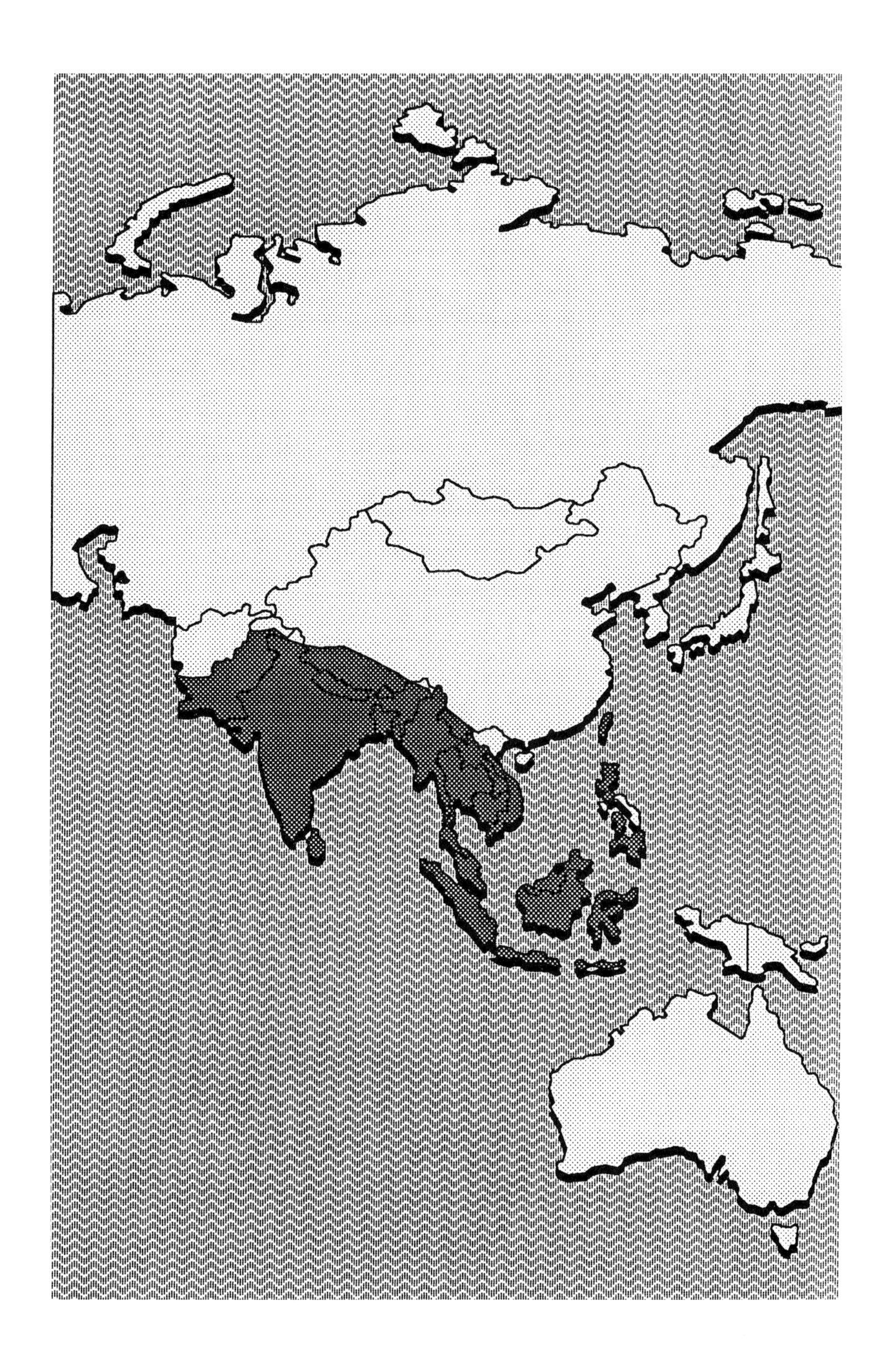

Egyptian cobra
Naja haje

Description: Yellowish, dark brown, or black uniform top with dark brown crossbands. Its head is sometimes black.

Characteristics: It is extremely dangerous. It is responsible for many human deaths. Once aroused or threatened, it will attack and continue to attack until it feels escape is possible. Its venom is neurotoxic and much stronger than the common cobra's. Its venom causes paralysis and death due to respiratory failure.

Habitat: Cultivated farmlands, open fields, and arid countrysides. It is often seen around homes searching for rodents.

Length: Average 1.5 meters, maximum 2.5 meters.

Distribution: Africa, Iraq, Syria, and Saudi Arabia.

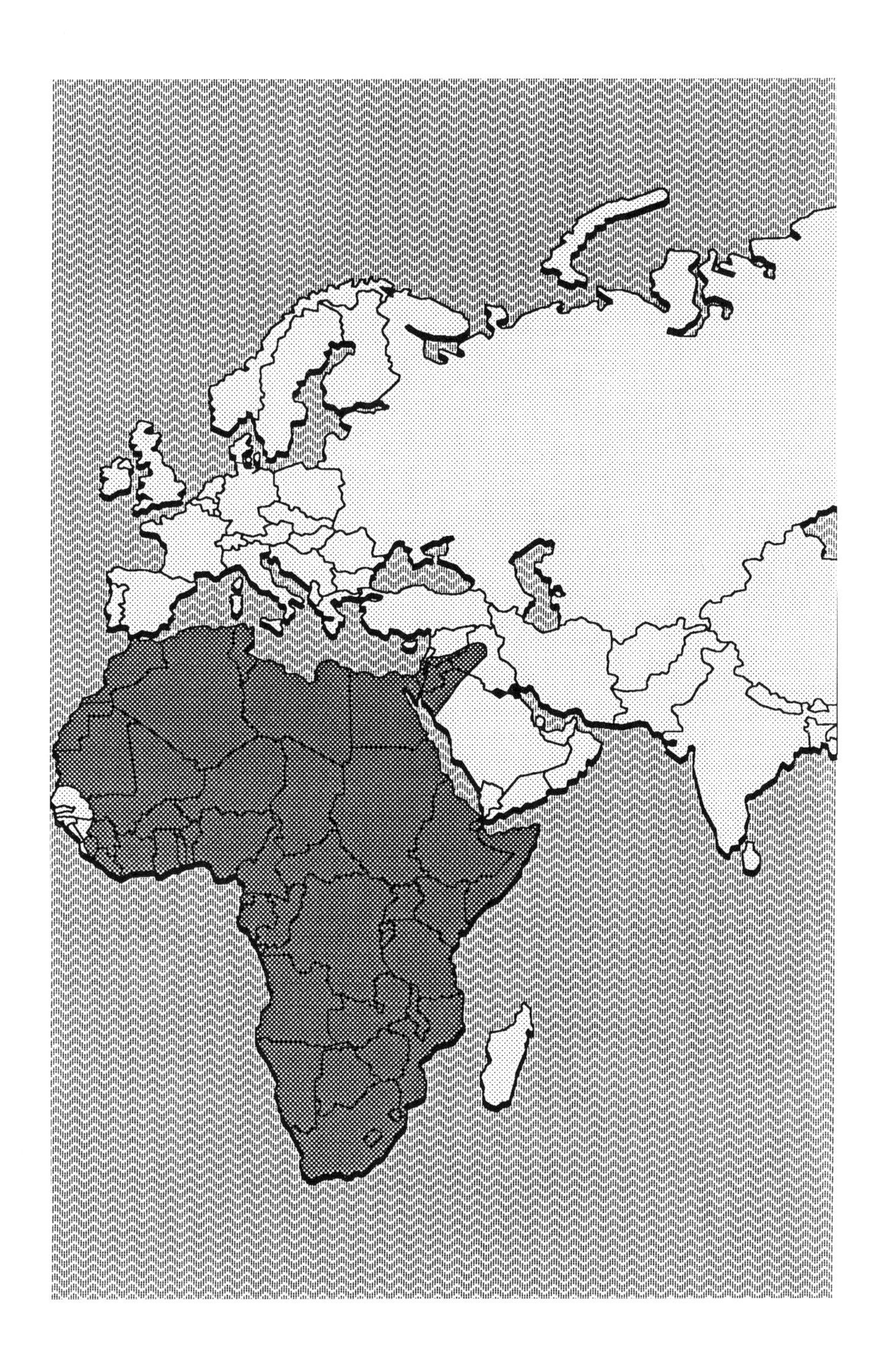

Gaboon viper

Bitis gabonica

Description: Pink to brown with a vertebral series of elongated yellowish or light brown spots connected by hourglass-shaped markings on each side. It has a dark brown stripe behind each eye. This dangerous viper is almost invisible on the forest floor. A 1.8-meter-long Gaboon viper could weigh 16 kilograms.

Characteristics: The largest and heaviest of all true vipers, having a very large triangular head. It comes out in the evening to feed. Fortunately, it is not aggressive, but it will stand its ground if approached. It bites when molested or stepped on. Its fangs are enormous, often measuring 5 centimeters long. It injects a large amount of venom when it strikes. Its venom is neurotoxic and hemotoxic.

Habitat: Dense rain forests. Occasionally found in open country.

Length: Average 1.2 meters, maximum 1.8 meters.

Distribution: Most of Africa.

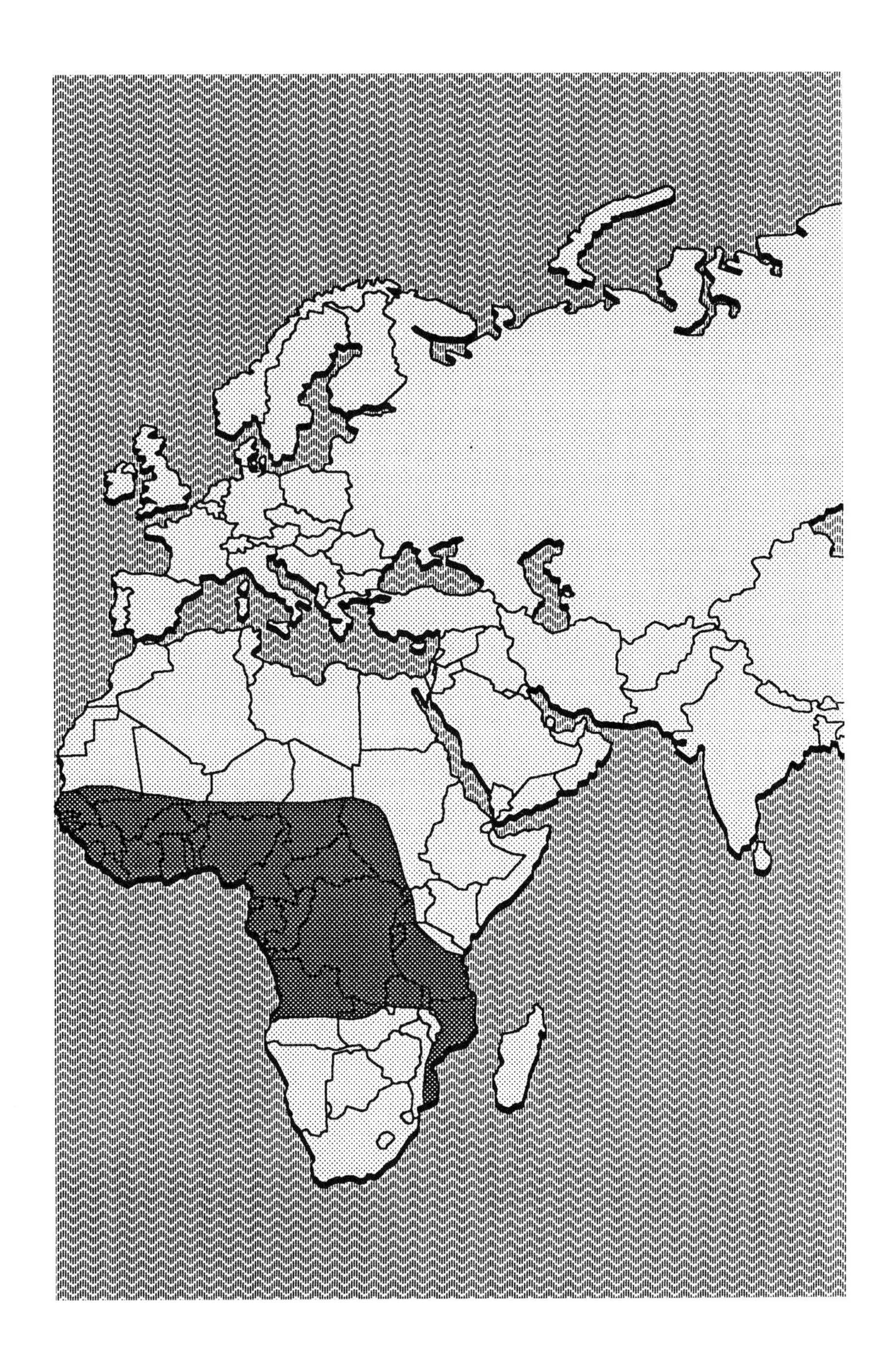

Green mamba

Dendraspis angusticeps

Description: Most mambas are uniformly bright green over their entire body. The black mamba, the largest of the species, is uniformly olive to black.

Characteristics: The mamba is the dreaded snake species of Africa. Treat it with great respect. It is considered one of the most dangerous snakes known. Not only is it highly venomous but it is aggressive and its victim has little chance to escape from a bite. Its venom is highly neurotoxic.

Habitat: Mambas are at home in the brush, trees, and low-hanging branches looking for birds, a usual diet for this species.

Length: Average 1.8 meters, maximum 3.7 meters.

Distribution: Most of Africa.

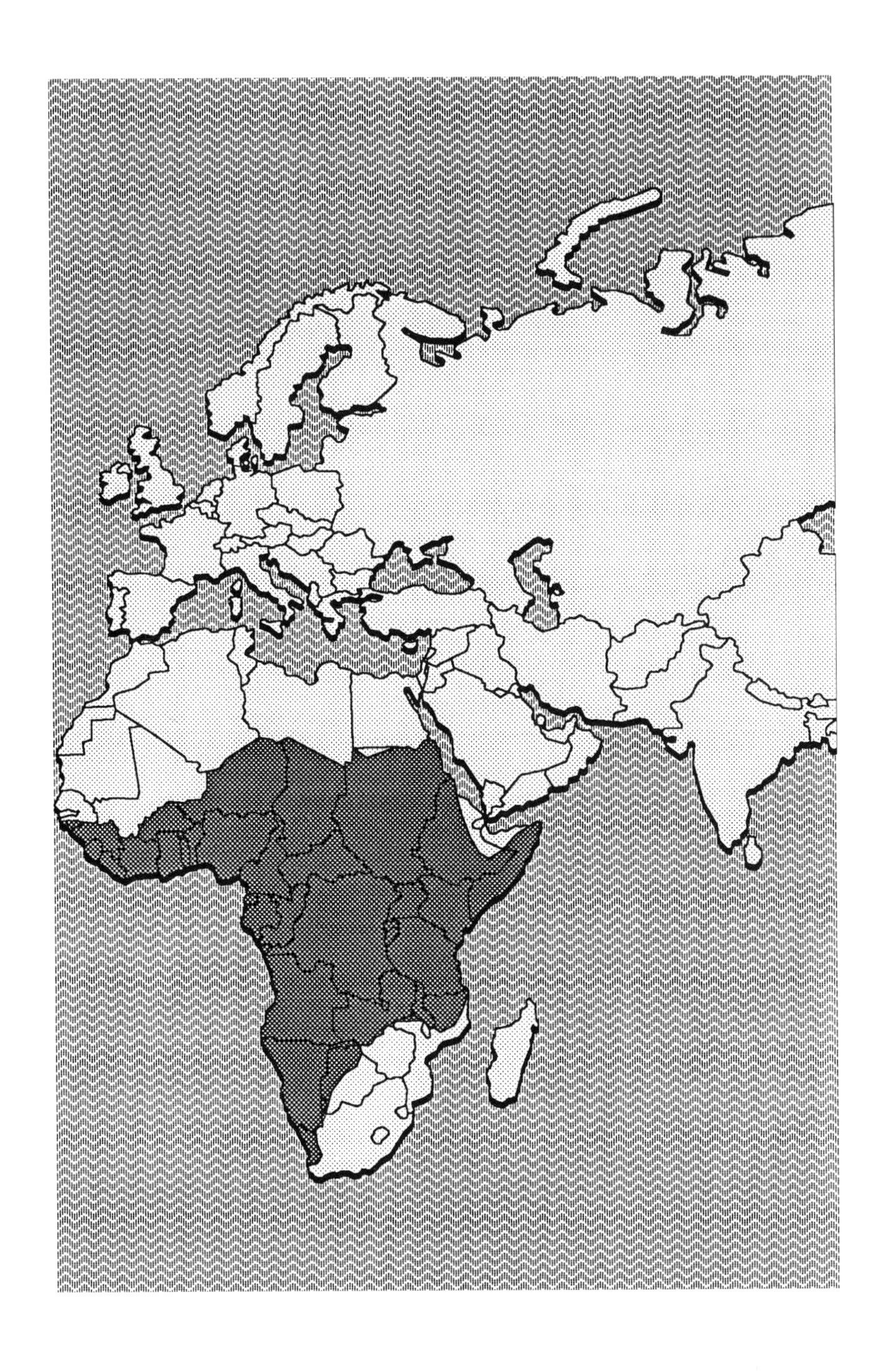

Green tree pit viper

Trimeresurus gramineus

Description: Uniform bright or dull green with light yellow on the facial lips.

Characteristics: A small arboreal snake of some importance, though not considered a deadly species. It is a dangerous species because most of its bites occur in the head, shoulder, and neck areas. It seldom comes to the ground. It feeds on young birds, lizards, and tree frogs.

Habitat: Found in dense rain forests and plantations.

Length: Average 45 centimeters, maximum 75 centimeters.

Distribution: India, Burma, Malaya, Thailand, Laos, Cambodia, Vietnam, China, Indonesia, and Formosa.

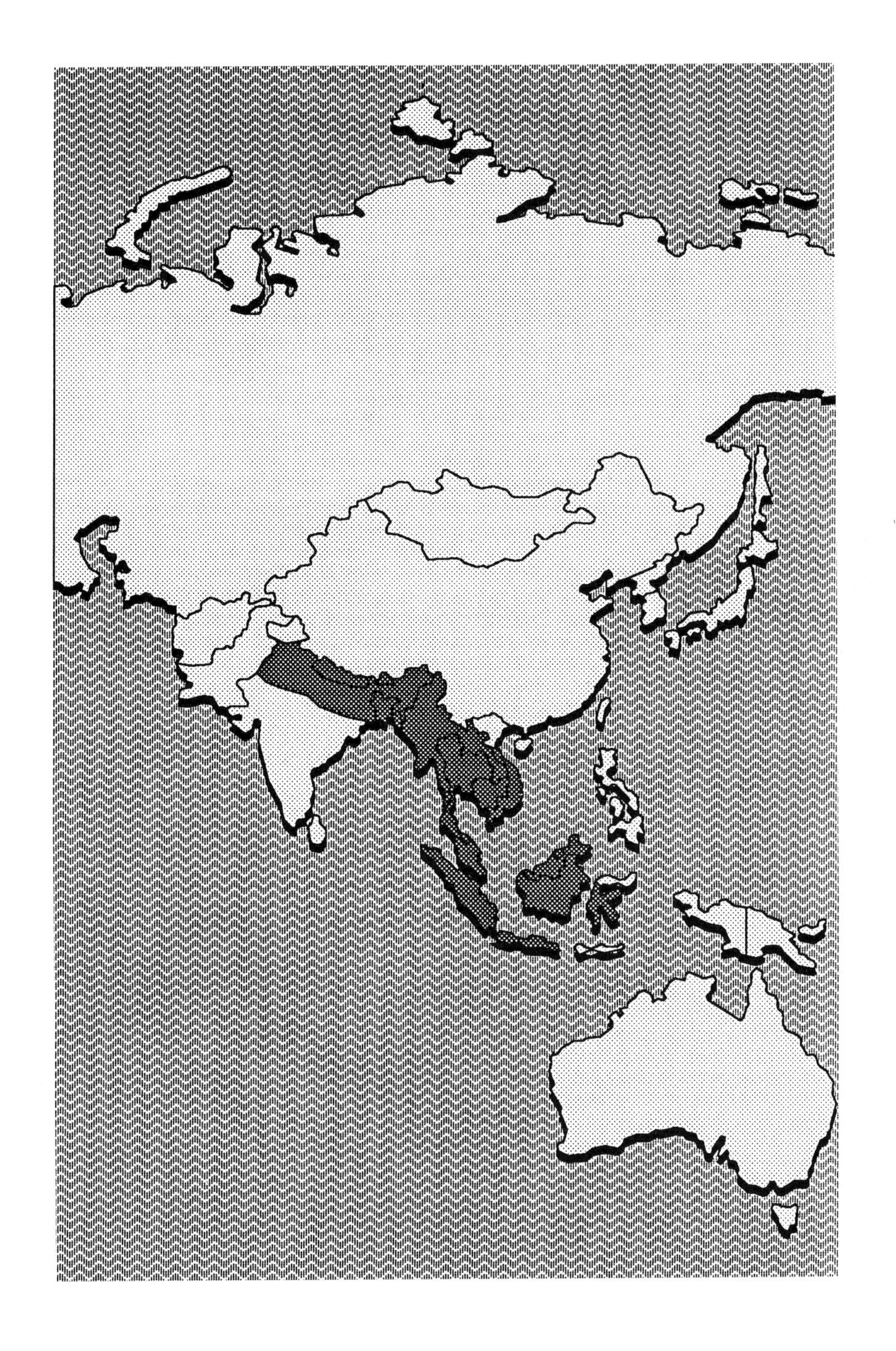

Habu pit viper

Trimeresurus flavoviridis

Description: Light brown or olive-yellow with black markings and a yellow or greenish-white belly.

Characteristics: This snake is responsible for biting many humans and its bite could be fatal. It is an irritable species ready to defend itself. Its venom is hemotoxic, causing pain and considerable tissue damage.

Habitat: Found in a variety of habitats, ranging from lowlands to mountainous regions. Often encountered in old houses and rock walls surrounding buildings.

Length: Average 1 meter, maximum 1.5 meters.

Distribution: Okinawa and neighboring islands and Kyushu.

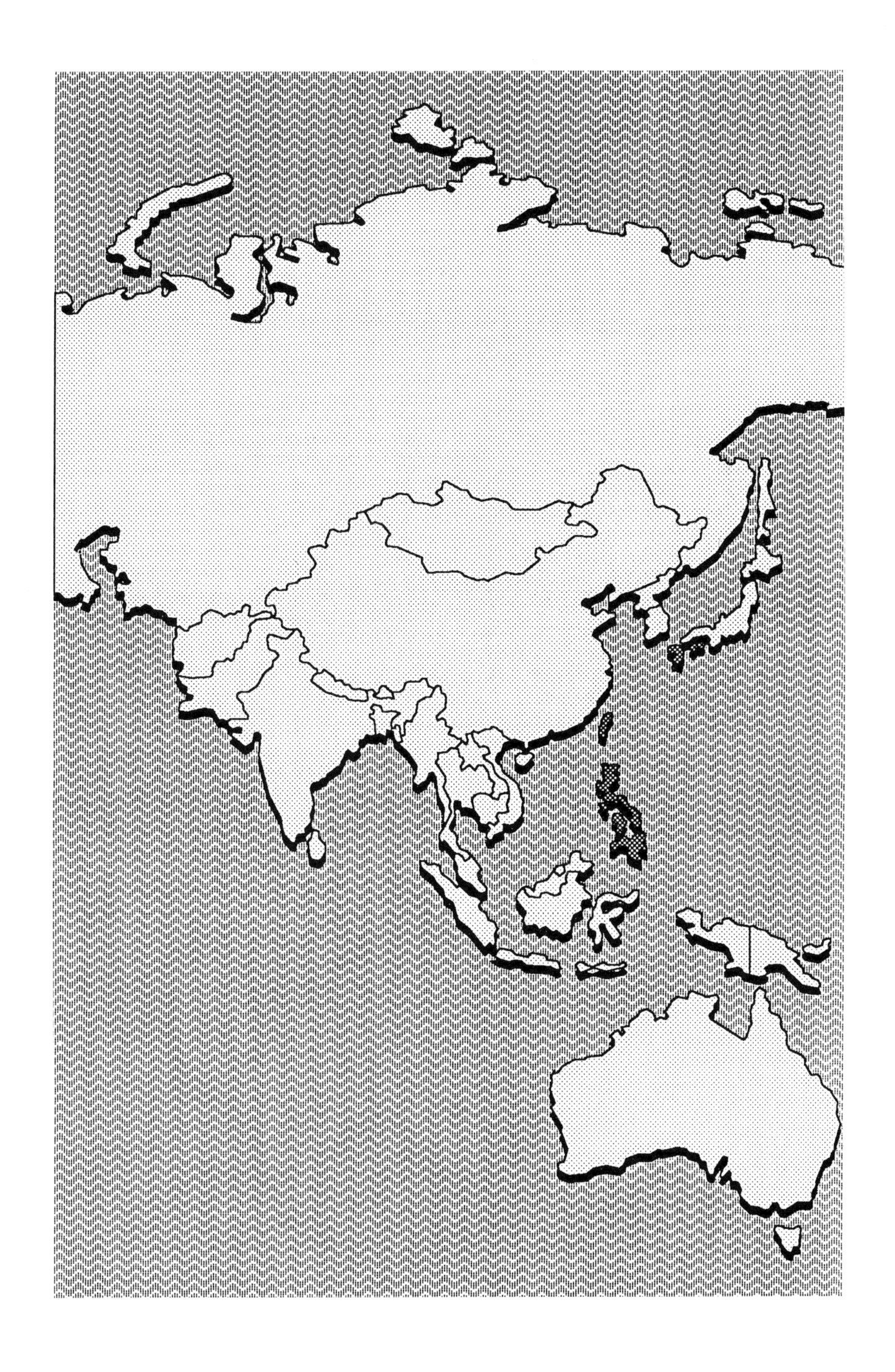

Horned desert viper

Cerastes cerastes

Description: Pale buff color with obscure markings and a sharp spine (scale) over each eye.

Characteristics: As with all true vipers that live in the desert, it finds refuge by burrowing in the heat of the day, coming out at night to feed. It is difficult to detect when buried; therefore, many bites result from the snake being accidentally stepped on. Its venom is hemotoxic, causing severe damage to blood cells and tissue.

Habitat: Only found in very arid places within its range.

Length: Average 45 centimeters, maximum 75 centimeters.

Distribution: Arabian Peninsula, Africa, Iran, and Iraq.

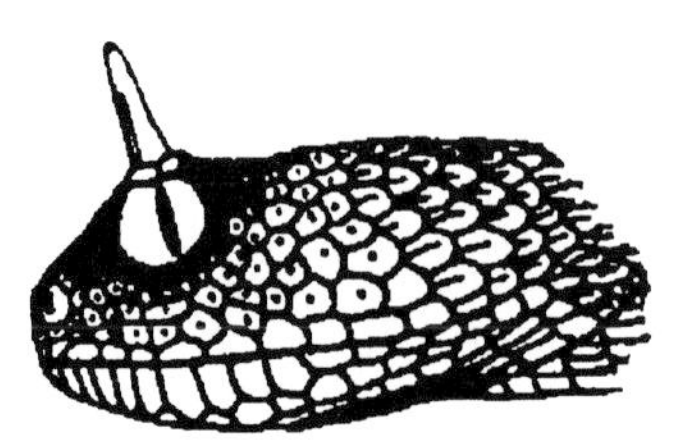

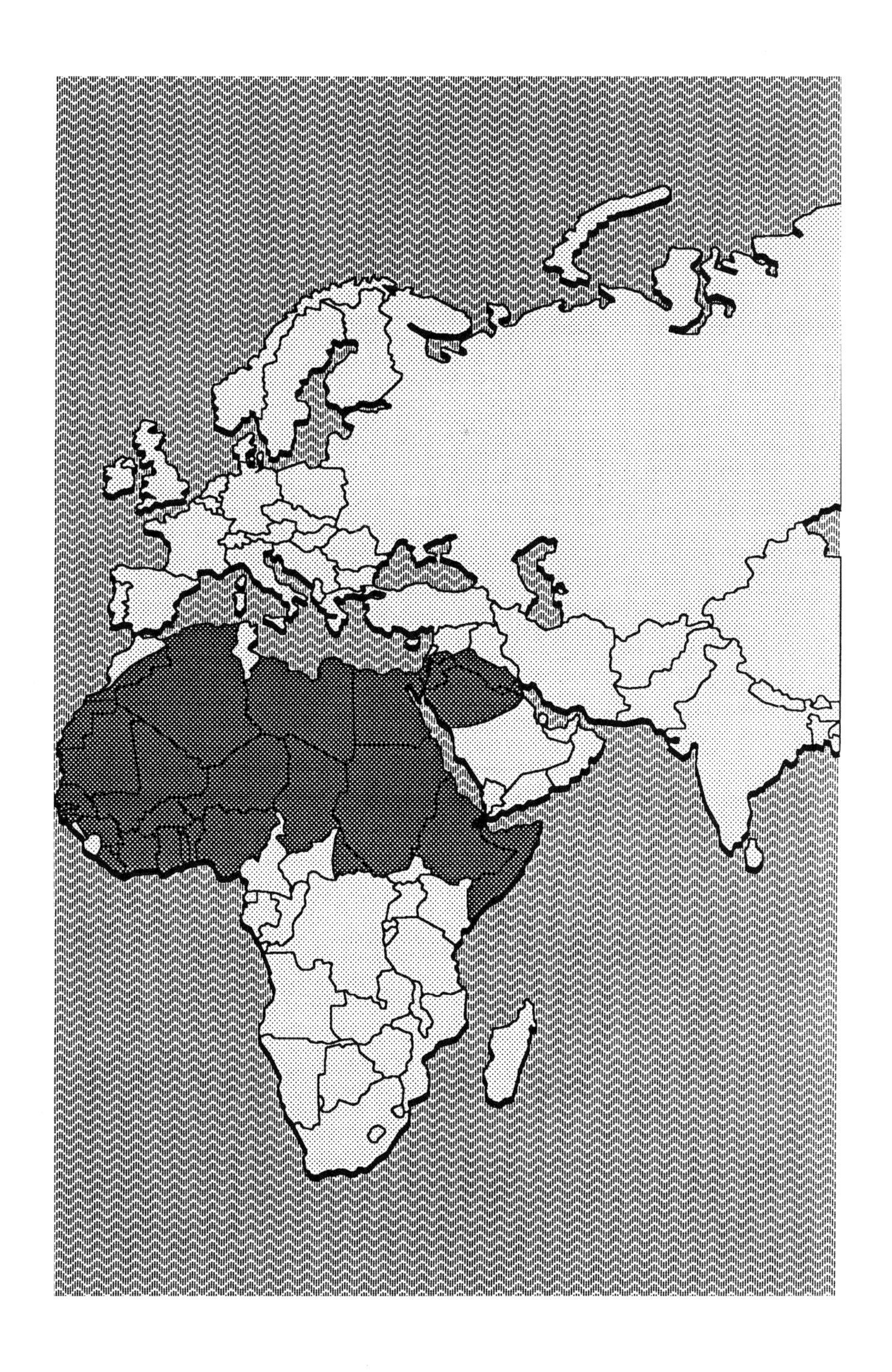

King cobra
Ophiophagus hannah

Description: Uniformly olive, brown, or green with ringlike crossbands of black.

Characteristics: Although it is the largest venomous snake in the world and it has a disposition to go with this honor, it causes relatively few bites on humans. It appears to have a degree of intelligence. It avoids attacking another venomous snake for fear of being bitten. It feeds exclusively on harmless species. The female builds a nest then deposits her eggs. Lying close by, she guards her nest and is highly aggressive toward anything that closely approaches the nest. Its venom is a powerful neurotoxin. Without medical aid, death is certain for its victims.

Habitat: Dense jungle and cultivated fields.

Length: Average 3.5 meters, maximum 5.5 meters.

Distribution: Thailand, southern China, Malaysia Peninsula, and the Philippines.

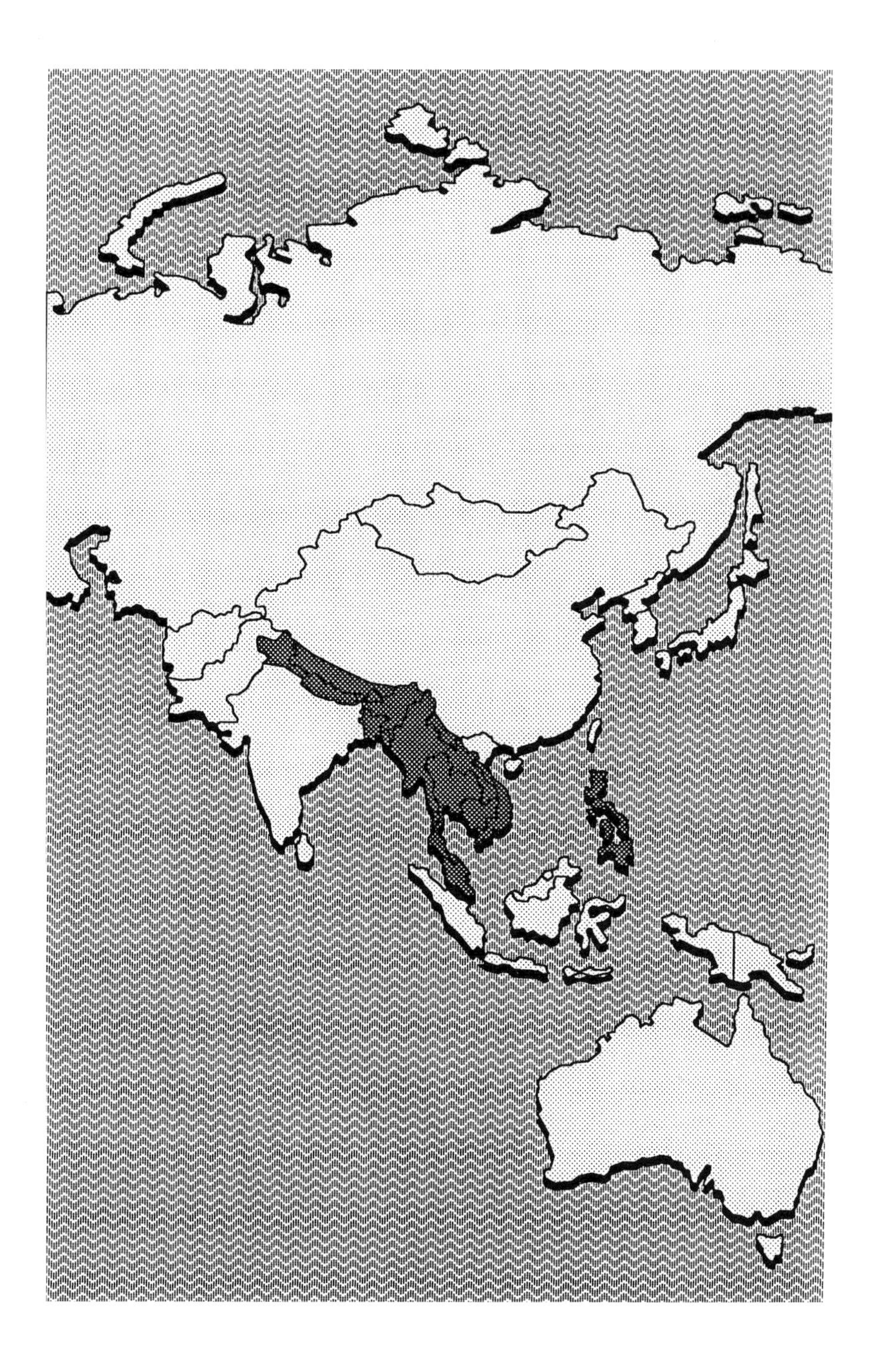

Krait

Bungarus caeruleus

Description: Black or bluish-black with white narrow crossbands and a narrow head.

Characteristics: Kraits are found only in Asia. This snake is of special concern to man. It is deadly – about 15 times more deadly than the common cobra. It is active at night and relatively passive during the day. The native people often step on kraits while walking through their habitats. The krait has a tendency to seek shelter in sleeping bags, boots, and tents. Its venom is a powerful neurotoxin that causes respiratory failure.

Habitat: Open fields, human settlements, and dense jungle.

Length: Average 90 centimeters, maximum 1.5 meters.

Distribution: India, Sri Lanka, and Pakistan.

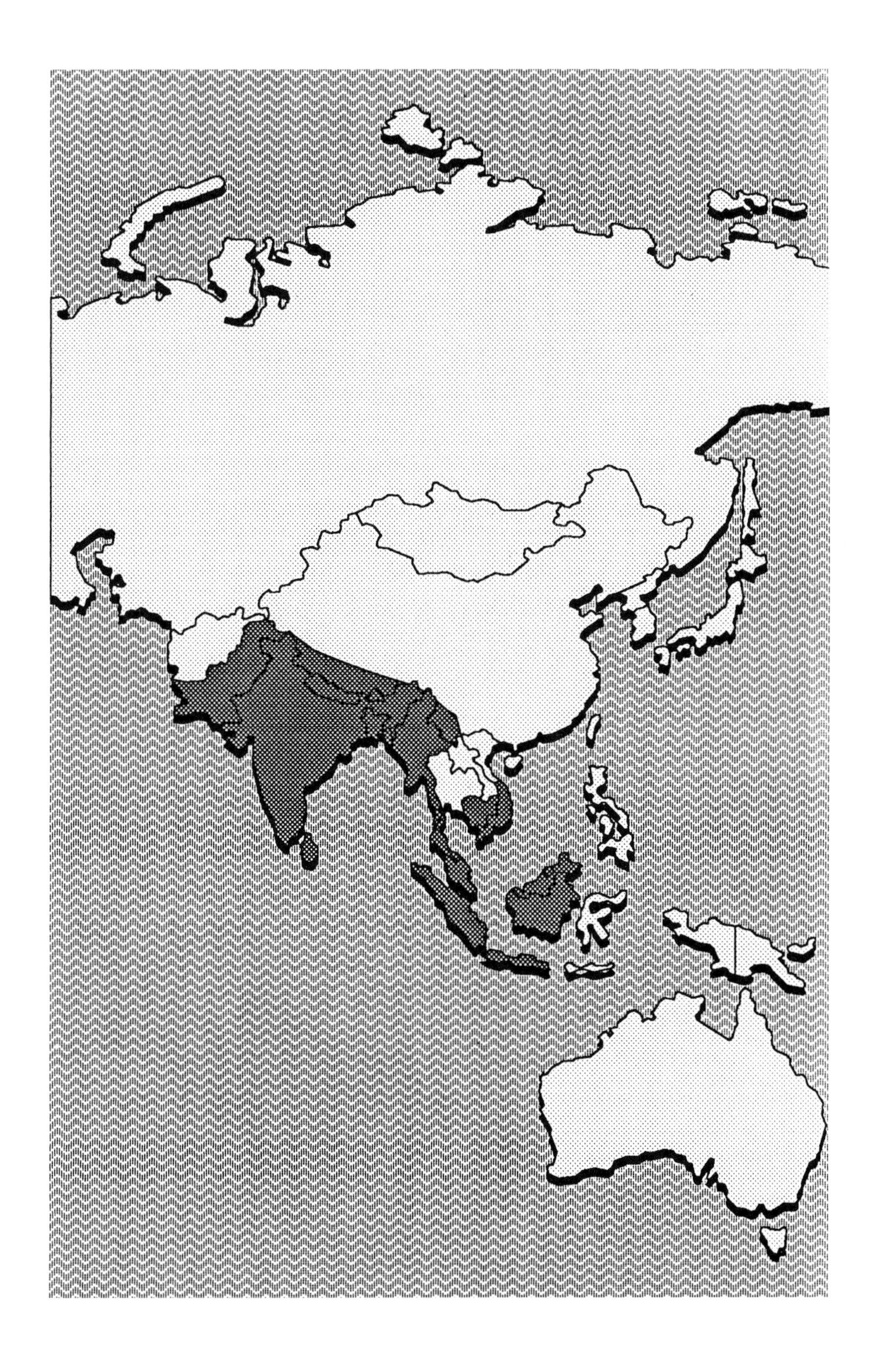

Levant viper

Vipera lebetina

Description: Gray to pale brown with large dark brown spots on the top of the back and a ^ mark on top of the head.

Characteristics: This viper belongs to a large group of true vipers. Like its cousins, it is large and dangerous. Its venom is hemotoxic. Many deaths have been reported from bites of this species. It is a strong snake with an irritable disposition; it hisses loudly when ready to strike.

Habitat: Varies greatly, from farmlands to mountainous areas.

Length: Average 1 meter, maximum 1.5 meters.

Distribution: Greece, Iraq, Syria, Lebanon, Turkey, Afghanistan, lower portion of the former USSR, and Saudi Arabia.

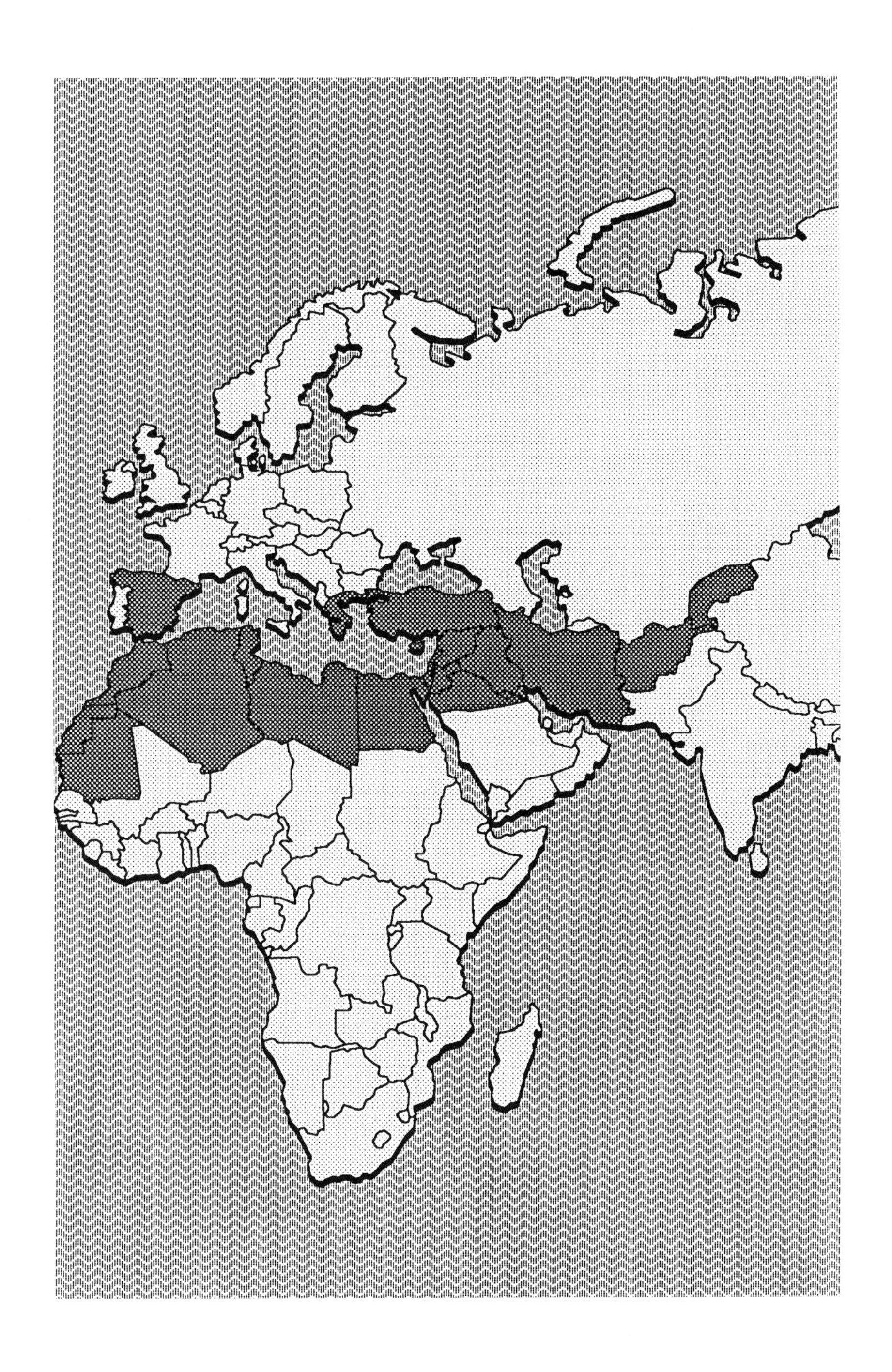

Malayan pit viper

Callaselasma rhodostoma

Description: Reddish running into pink tinge toward the belly with triangular-shaped, brown markings bordered with light-colored scales. The base of the triangular-shaped markings ends at the midline. It has dark brown, arrow-shaped markings on the top and each side of its head.

Characteristics: This snake has long fangs, is ill-tempered, and is responsible for many bites. Its venom is hemotoxic, destroying blood cells and tissue, but a victim's chances of survival are good with medical aid. This viper is a ground dweller that moves into many areas in search of food. The greatest danger is in stepping on the snake with bare feet.

Habitat: Rubber plantations, farms, rural villages, and rain forests.

Length: Average 60 centimeters, maximum 1 meter.

Distribution: Thailand, Laos, Cambodia, Java, Sumatra, Malaysia, Vietnam, Burma, and China.

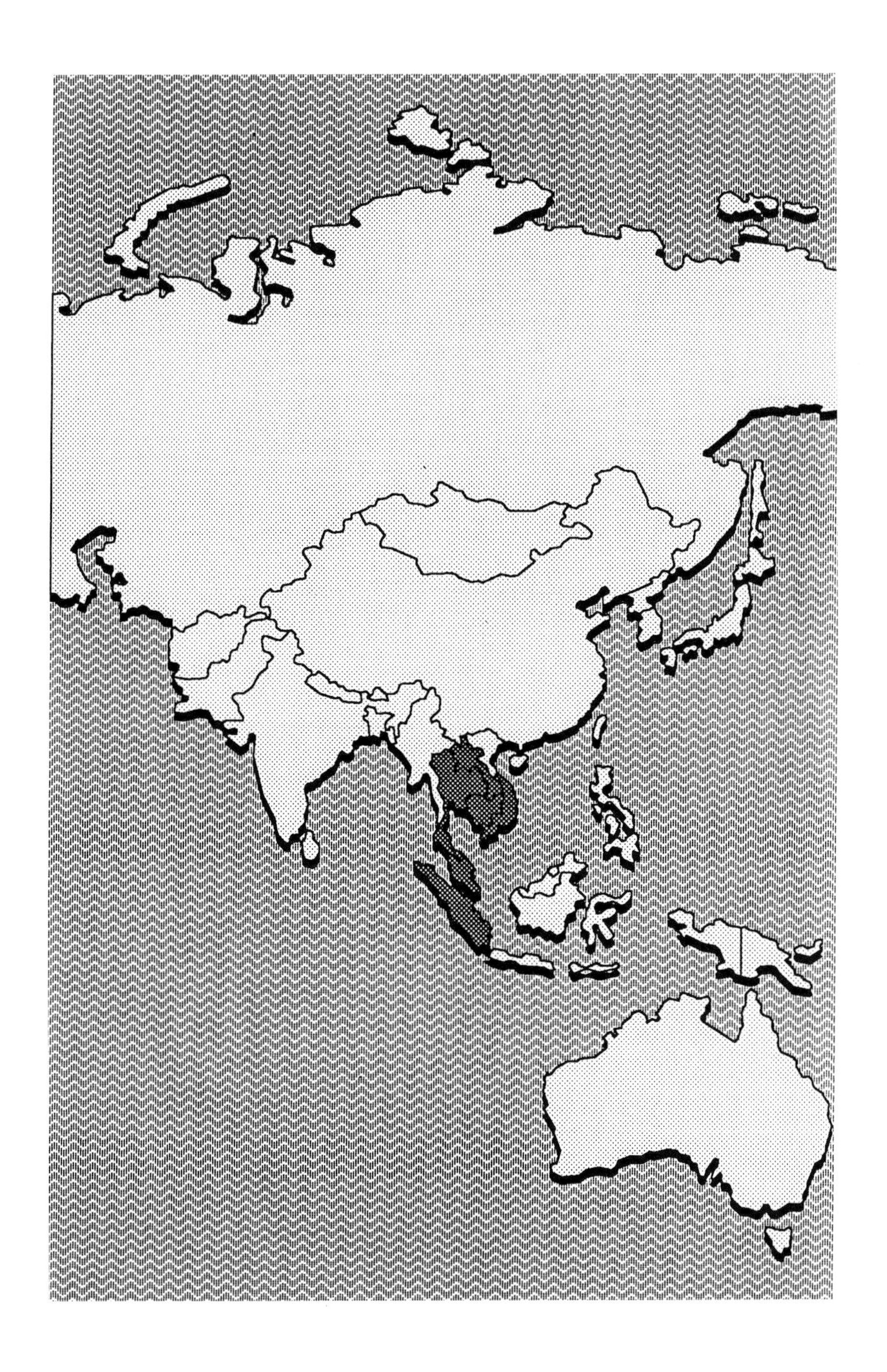

McMahon's viper

Eristicophis macmahonii

Description: Sandy buff color dominates the body with darker brown spots on the side of the body. Its nose shield is broad, aiding in burrowing.

Characteristics: Very little is known about this species. It apparently is rare or seldom seen. This viper is very irritable; it hisses, coils, and strikes at any intruder that ventures too close. Its venom is highly hemotoxic, causing great pain and tissue damage.

Habitat: Arid or semidesert. It hides during the day's sun, coming out only at night to feed on rodents.

Length: Average 45 centimeters, maximum 1 meter.

Distribution: West Pakistan, Afghanistan, and Iran.

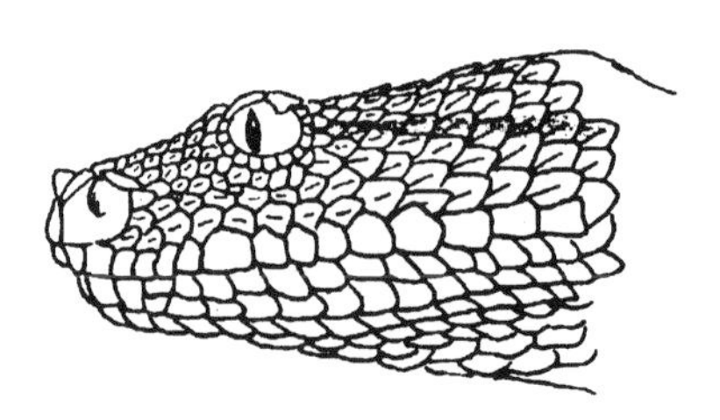

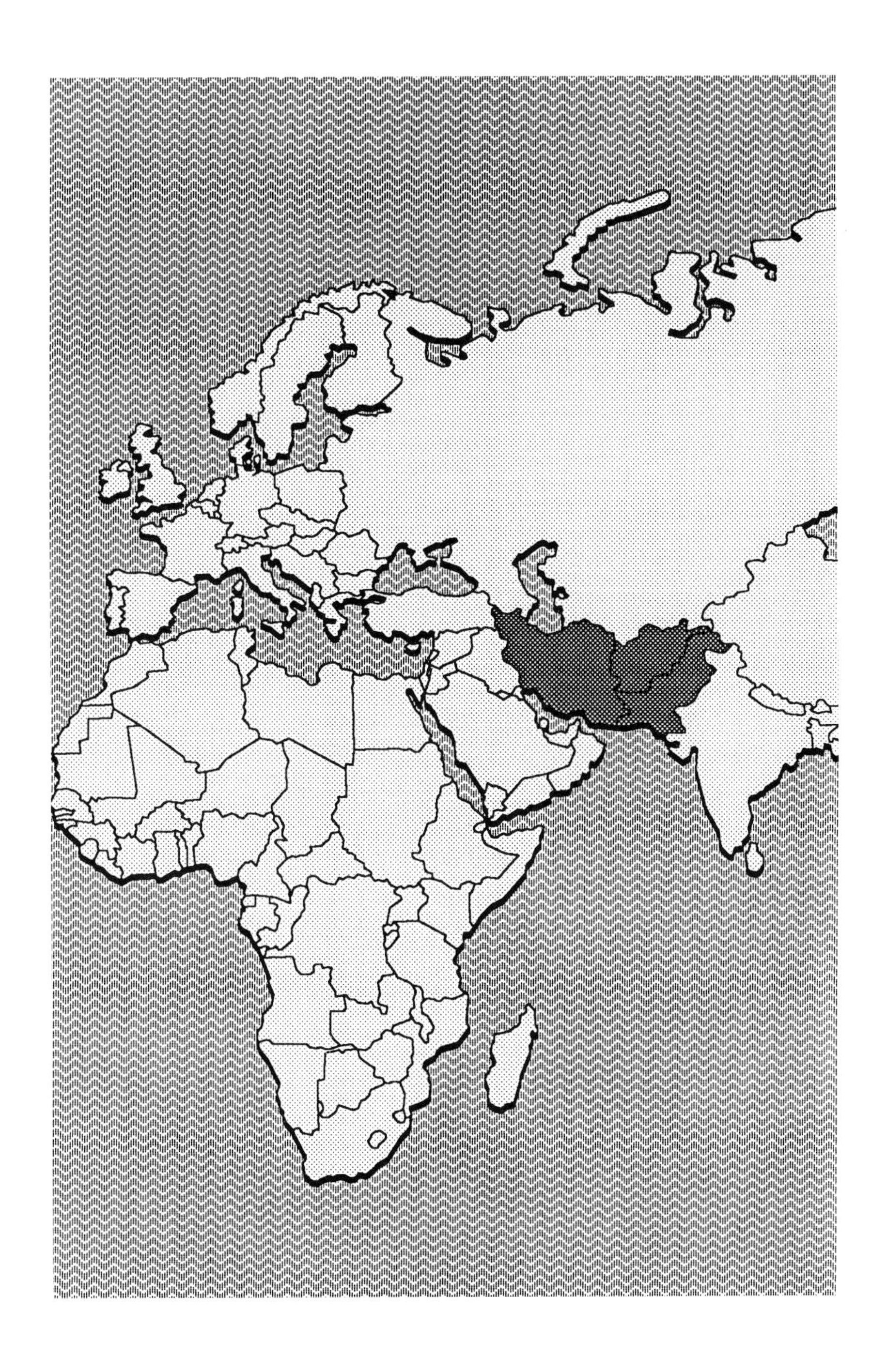

Mole viper or burrowing viper

Atracaspis microlepidota

Description: Uniformly black or dark brown with a small, narrow head.

Characteristics: A viper that does not look like one. It is small in size, and its small head does not indicate the presence of venom glands. It has a rather inoffensive disposition; however, it will quickly turn and bite if restrained or touched. Its venom is a potent hemotoxin for such a small snake. Its fangs are exceptionally long. A bite can result even when picking it up behind the head. It is best to leave this snake alone.

Habitat: Agricultural areas and arid localities.

Length: Average 55 centimeters, maximum 75 centimeters.

Distribution: Sudan, Ethiopia, Somaliland, Kenya, Tanganyika, Uganda, Cameroon, Niger, Congo, and Urundi.

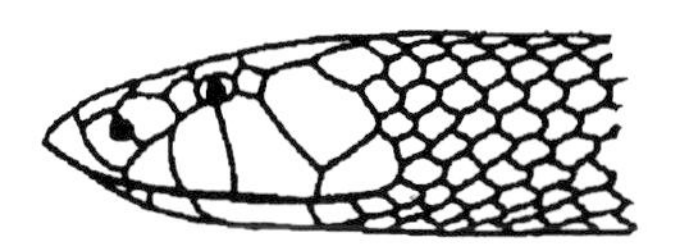

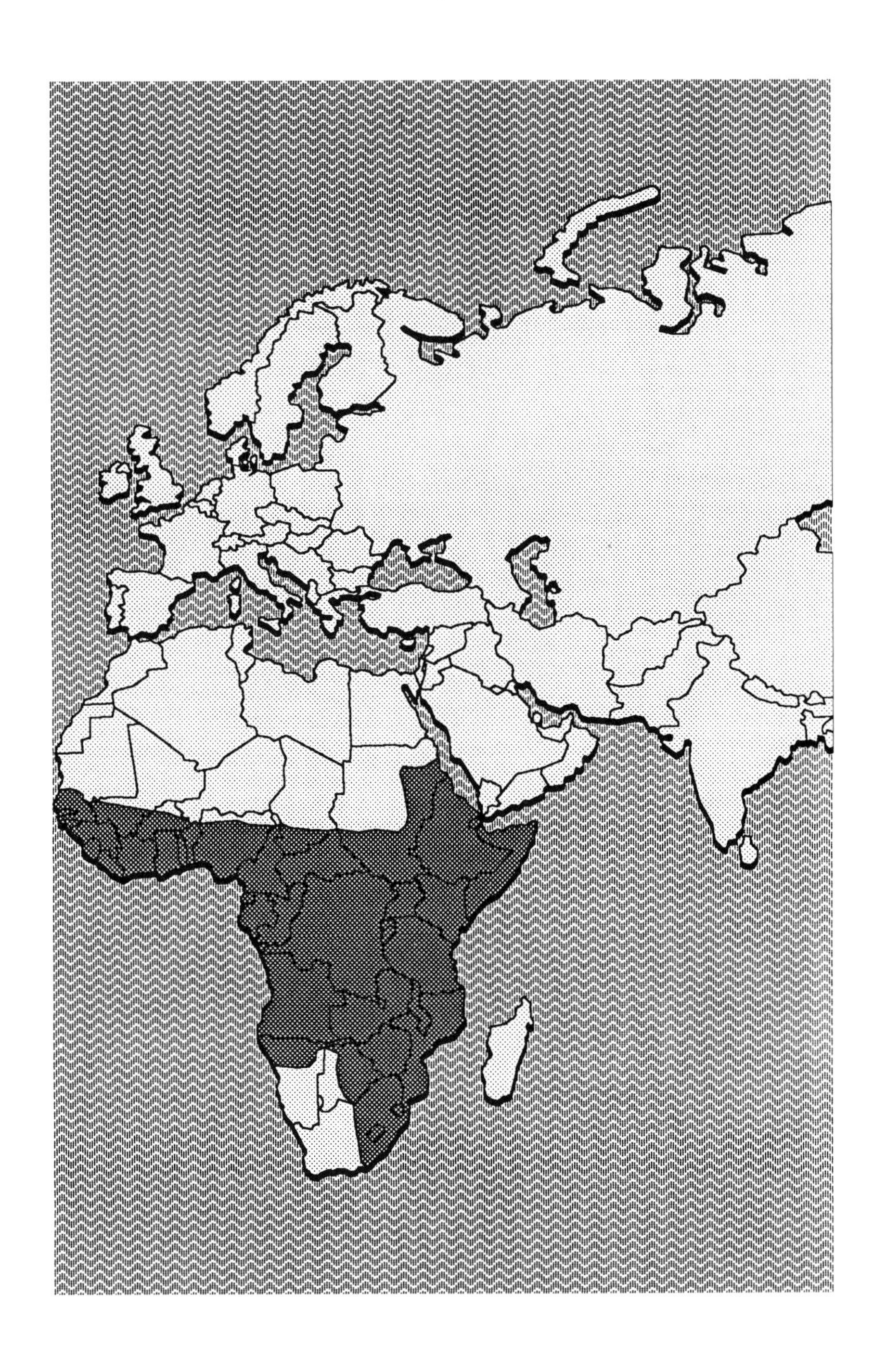

Palestinian viper

Vipera palaestinae

Description: Olive to rusty brown with a dark V-shaped mark on the head and a brown, zigzag band along the back.

Characteristics: The Palestinian viper is closely related to the Russell's viper of Asia. Like its cousin, it is extremely dangerous. It is active and aggressive at night but fairly placid during the day. When threatened or molested, it will tighten its coils, hiss loudly, and strike quickly.

Habitat: Arid regions, but may be found around barns and stables. It has been seen entering houses in search of rodents.

Length: Average 0.8 meter, maximum 1.3 meters.

Distribution: Turkey, Syria, Palestine, Israel, Lebanon, and Jordan.

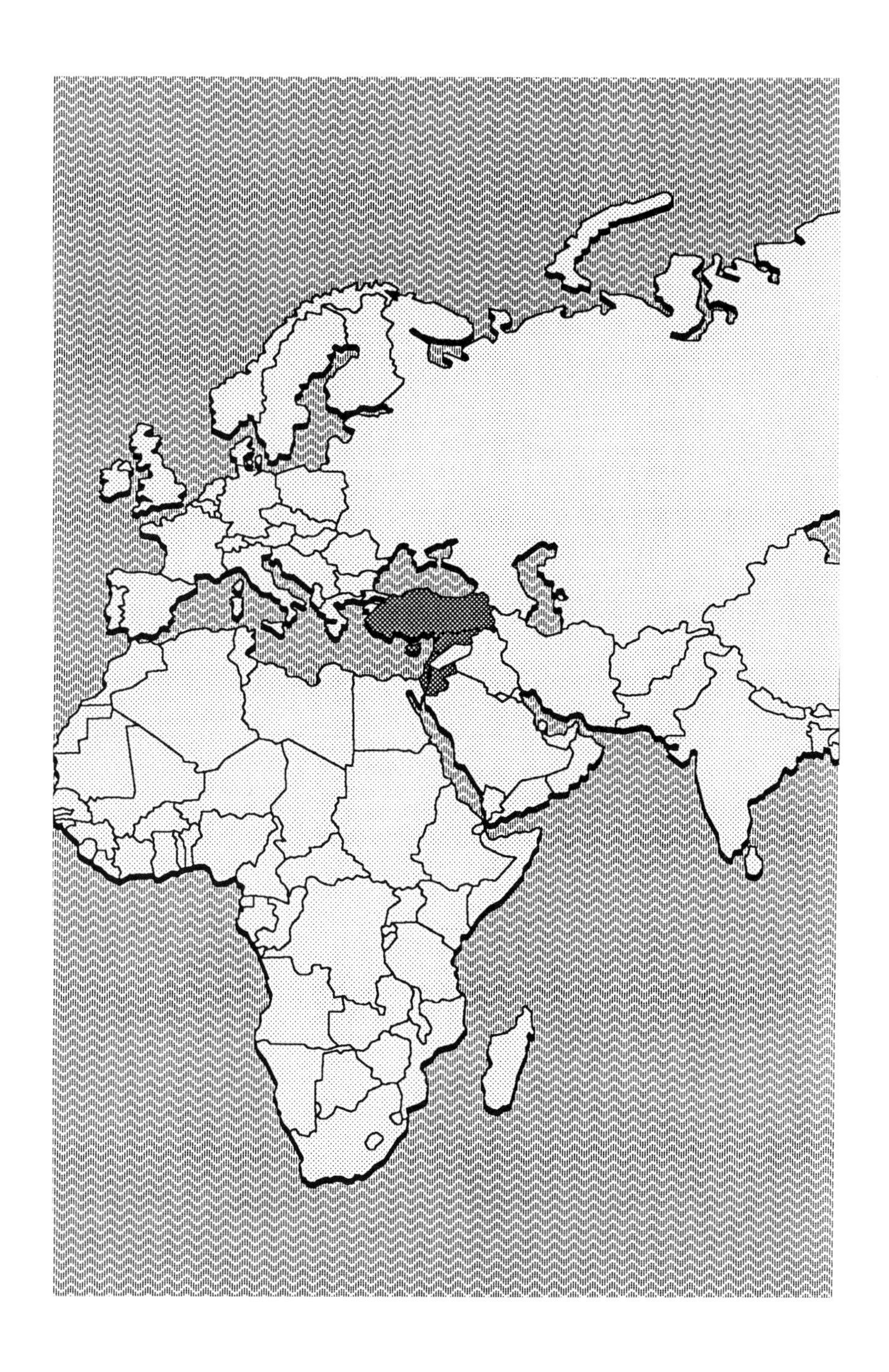

Puff adder

Bitis arietans

Description: Yellowish, light brown, or orange with chevron-shaped dark brown or black bars.

Characteristics: The puff adder is the second largest of the dangerous vipers. It is one of the most common snakes in Africa. It is largely nocturnal, hunting at night and seeking shelter during the day's heat. It is not shy when approached. It draws its head close to its coils, makes a loud hissing sound, and is quick to strike any intruder. Its venom is strongly hemotoxic, destroying blood cells and causing extensive tissue damage.

Habitat: Arid regions to swamps and dense forests. Common around human settlements.

Length: Average 1.2 meters, maximum 1.8 meters.

Distribution: Most of Africa, Saudi Arabia, Iraq, Lebanon, Israel, and Jordan.

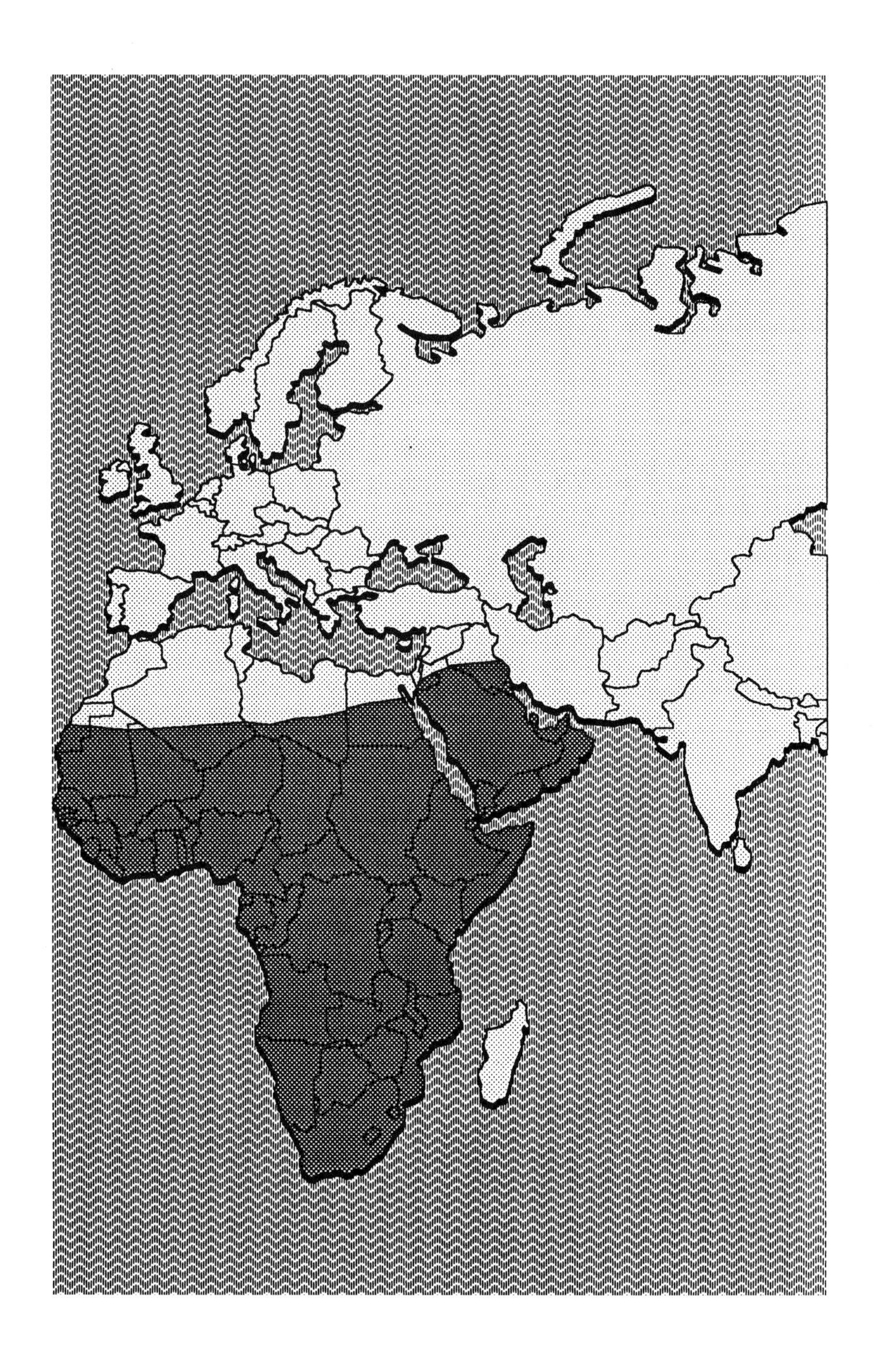

Rhinoceros viper or river jack

Bitis nasicornis

Description: Brightly colored with purplish to reddish-brown markings and black and light olive markings along the back. On its head it has a triangular marking that starts at the tip of the nose. It has a pair of long horns (scales) on the tip of its nose.

Characteristics: Its appearance is awesome; its horns and very rough scales give it a sinister look. It has an irritable disposition. It is not aggressive but will stand its ground ready to strike if disturbed. Its venom is neurotoxic and hemotoxic.

Habitat: Rain forests, along waterways, and in swamps.

Length: Average 75 centimeters, maximum 1 meter.

Distribution: Equatorial Africa.

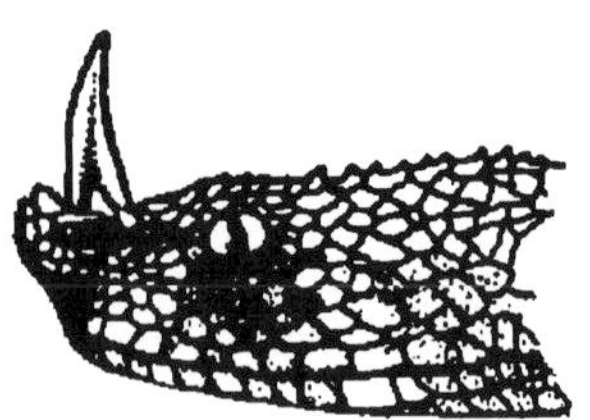

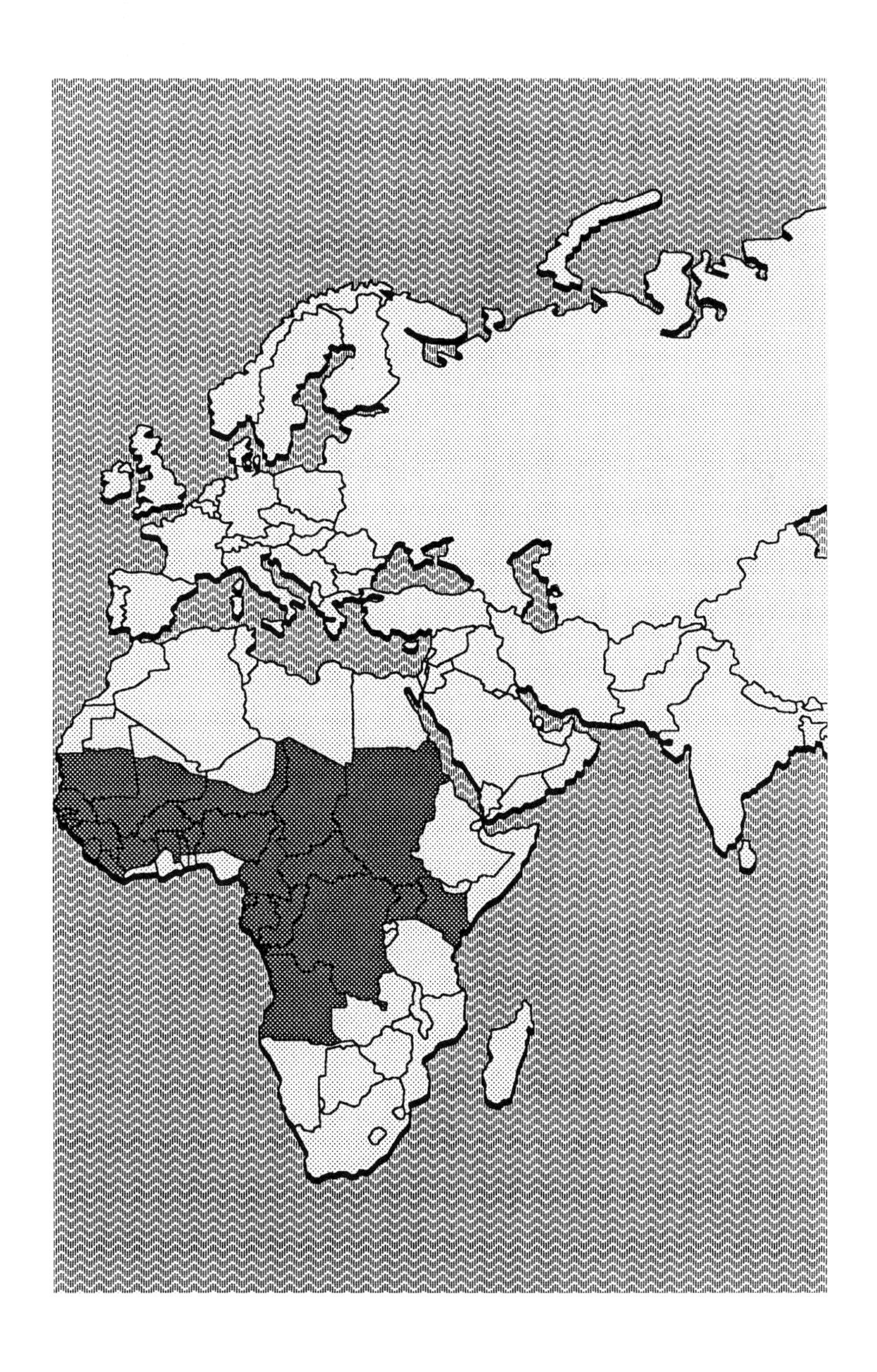

Russell's viper

Vipera russellii

Description: Light brown body with three rows of dark brown or black splotches bordered with white or yellow extending its entire length.

Characteristics: This dangerous species is abundant over its entire range. It is responsible for more human fatalities than any other venomous snake. It is irritable. When threatened, it coils tightly, hisses, and strikes with such speed that its victim has little chance of escaping. Its hemotoxic venom is a powerful coagulant, damaging tissue and blood cells.

Habitat: Variable, from farmlands to dense rain forests. It is commonly found around human settlements.

Length: Average 1 meter, maximum 1.5 meters.

Distribution: Sri Lanka, south China, India, Malaysian Peninsula, Java, Sumatra, Borneo, and surrounding islands.

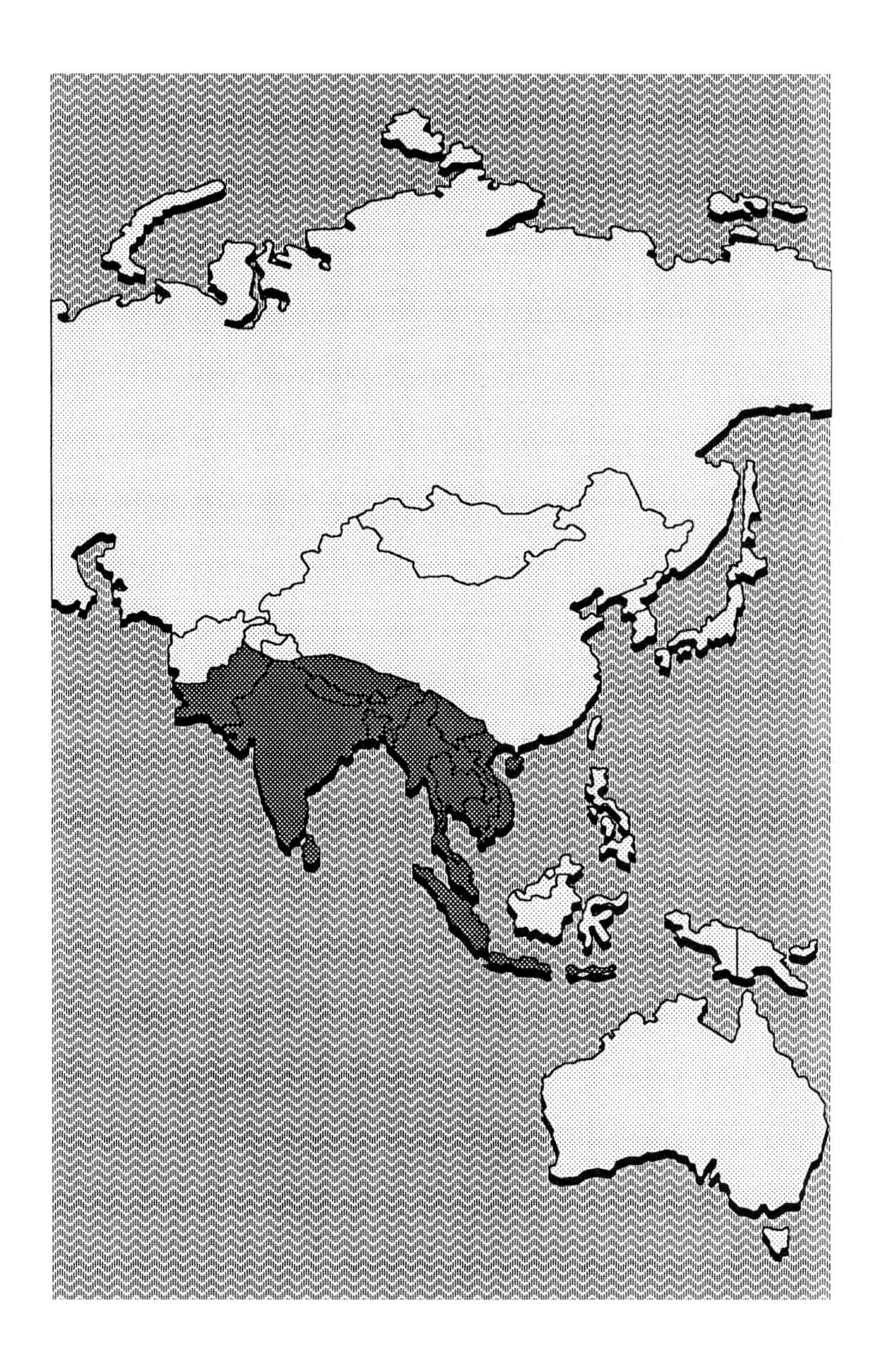

Sand viper

Cerastes vipera

Description: Usually uniformly very pallid, with three rows of darker brown spots.

Characteristics: A very small desert dweller that can bury itself in the sand during the day's heat. It is nocturnal, coming out at night to feed on lizards and small desert rodents. It has a short temper and will strike several times. Its venom is hemotoxic.

Habitat: Restricted to desert areas.

Length: Average 45 centimeters, maximum 60 centimeters.

Distribution: Northern Sahara, Algeria, Egypt, Sudan, Nigeria, Chad, Somalia, and central Africa.

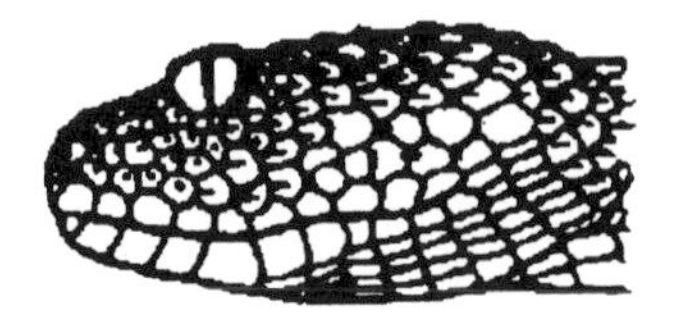

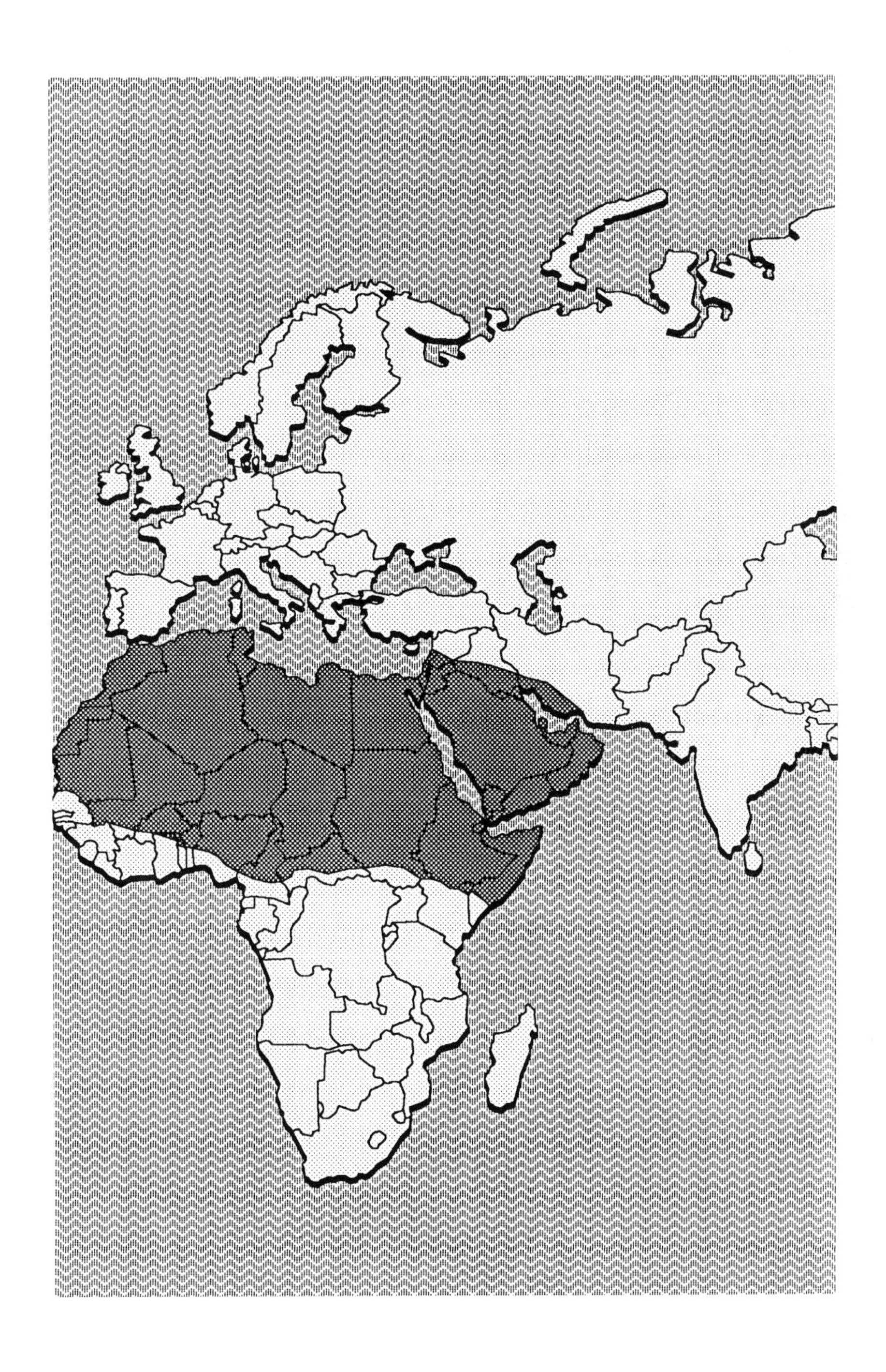

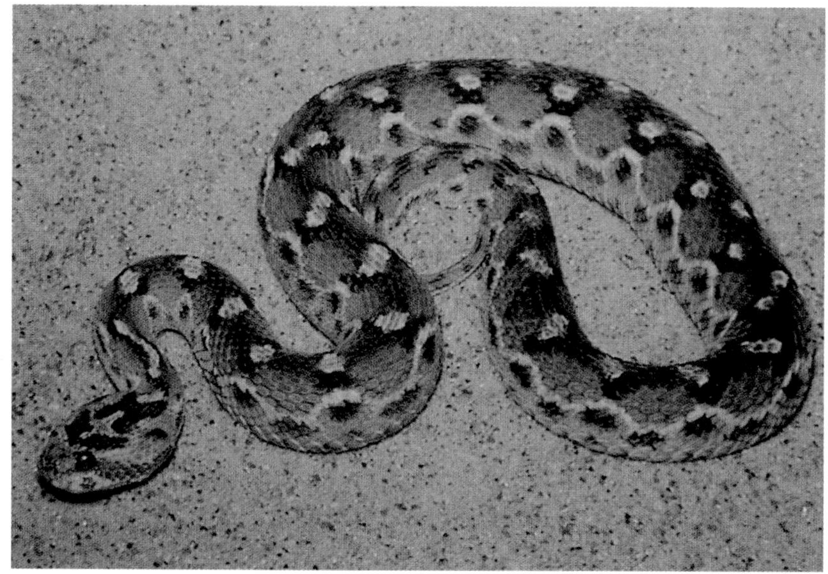

JOHN H. TASHJIAN/FORT WORTH ZOO

Saw-scaled viper

Echis carinatus

Description: Color is light buff with shades of brown, dull red, or gray. Its sides have a white or light-colored pattern. Its head usually has two dark stripes that start behind the eye and extend to the rear.

Characteristics: A small but extremely dangerous viper. It gets the name saw-scaled from rubbing the sides of its body together, producing a rasping sound. This ill-tempered snake will attack any intruder. Its venom is highly hemotoxic and quite potent. Many deaths are attributed to this species.

Habitat: Found in a variety of environments. It is common in rural settlements, cultivated fields, arid regions, barns, and rock walls.

Length: Average 45 centimeters, maximum 60 centimeters.

Distribution: Asia, Syria, India, Africa, Iraq, Iran, Saudi Arabia, Pakistan, Jordan, Lebanon, Sri Lanka, Algeria, Egypt, and Israel.

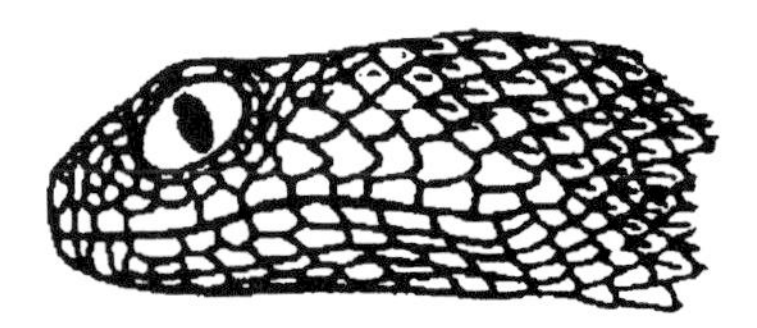

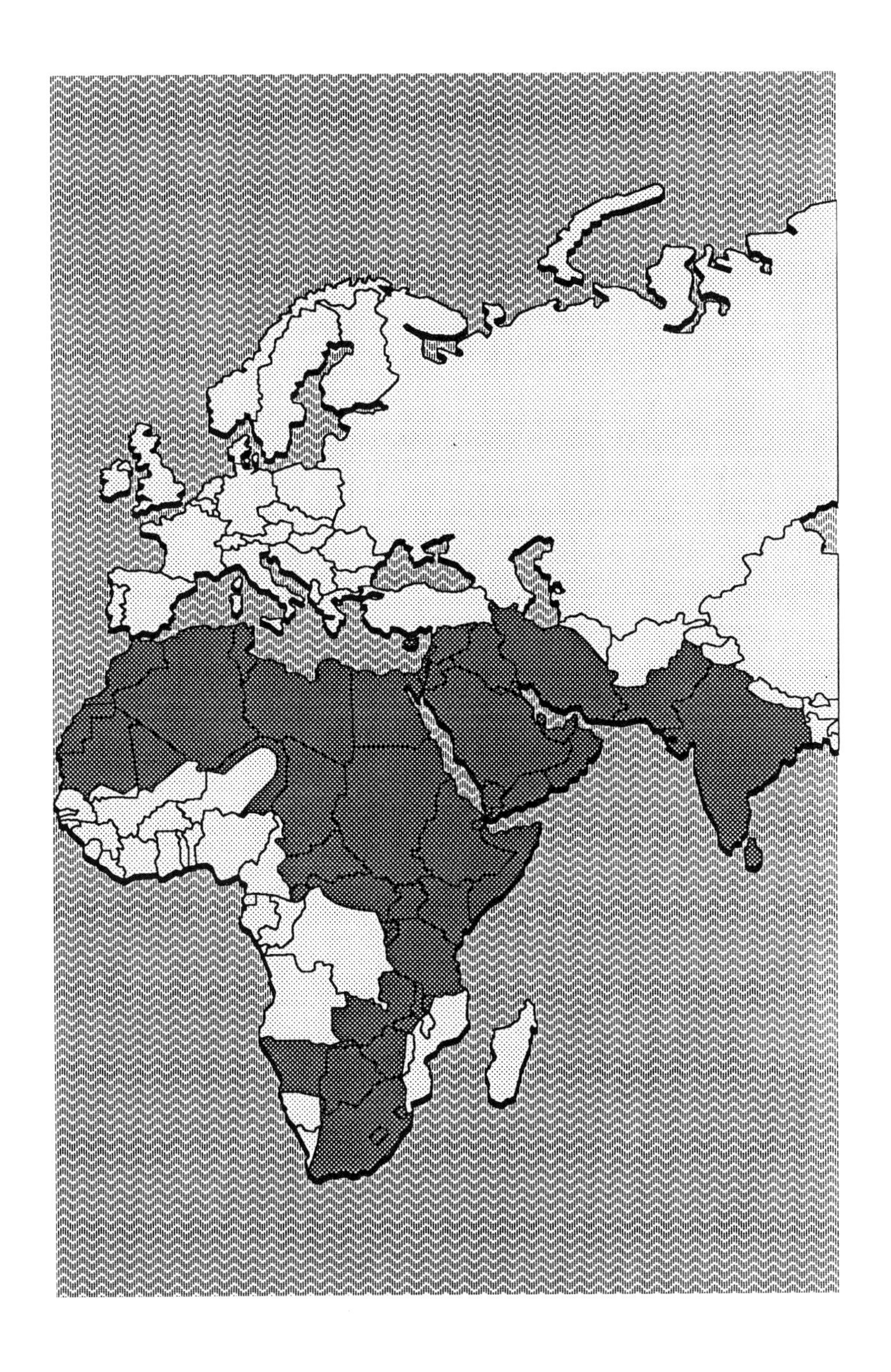

Wagler's pit viper or temple viper

Trimeresurus wagleri

Description: Green with white crossbands edged with blue or purple. It has two dorsal lines on both sides of its head.

Characteristics: It is also known as the temple viper because certain religious cults have placed venomous snakes in their temples. Bites are not uncommon for the species; fortunately, fatalities are very rare. It has long fangs. Its venom is hemotoxic causing cell and tissue destruction. It is an arboreal species and its bites often occur on the upper extremities.

Habitat: Dense rain forests, but often found near human settlements.

Length: Average 60 centimeters, maximum 100 centimeters.

Distribution: Malaysian Peninsula and Archipelago, Indonesia, Borneo, the Philippines, and Ryuku Islands.

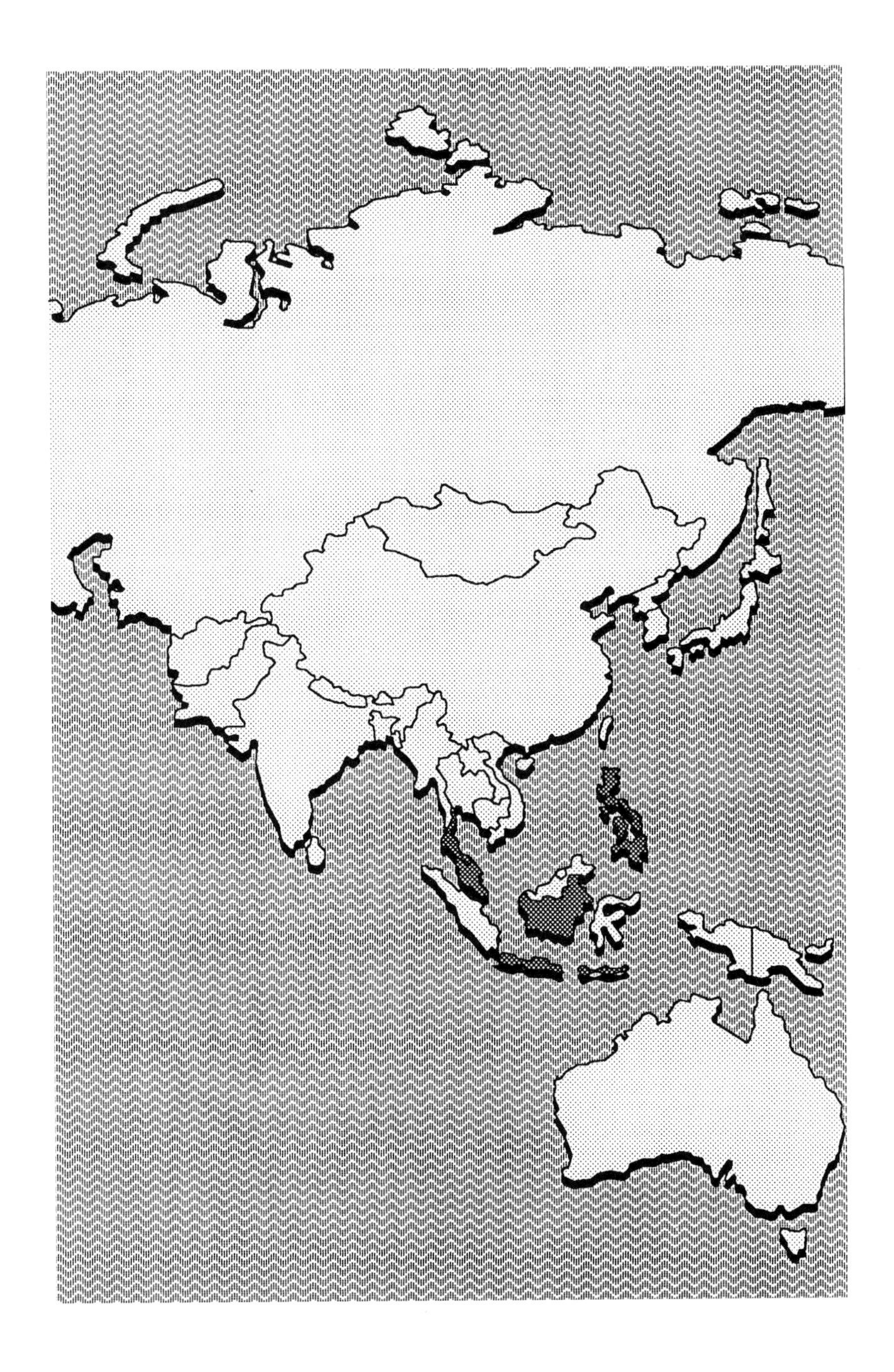

POISONOUS SNAKES OF AUSTRALASIA

Australian copperhead

Denisonia superba

Description: Coloration is reddish brown to dark brown. A few from Queensland are black.

Characteristics: Rather sluggish disposition but will bite if stepped on. When angry, rears its head a few inches from the ground with its neck slightly arched. Its venom is neurotoxic.

Habitat: Swamps.

Length: Average 1.2 meters, maximum 1.8 meters.

Distribution: Tasmania, South Australia, Queensland, and Kangaroo Island.

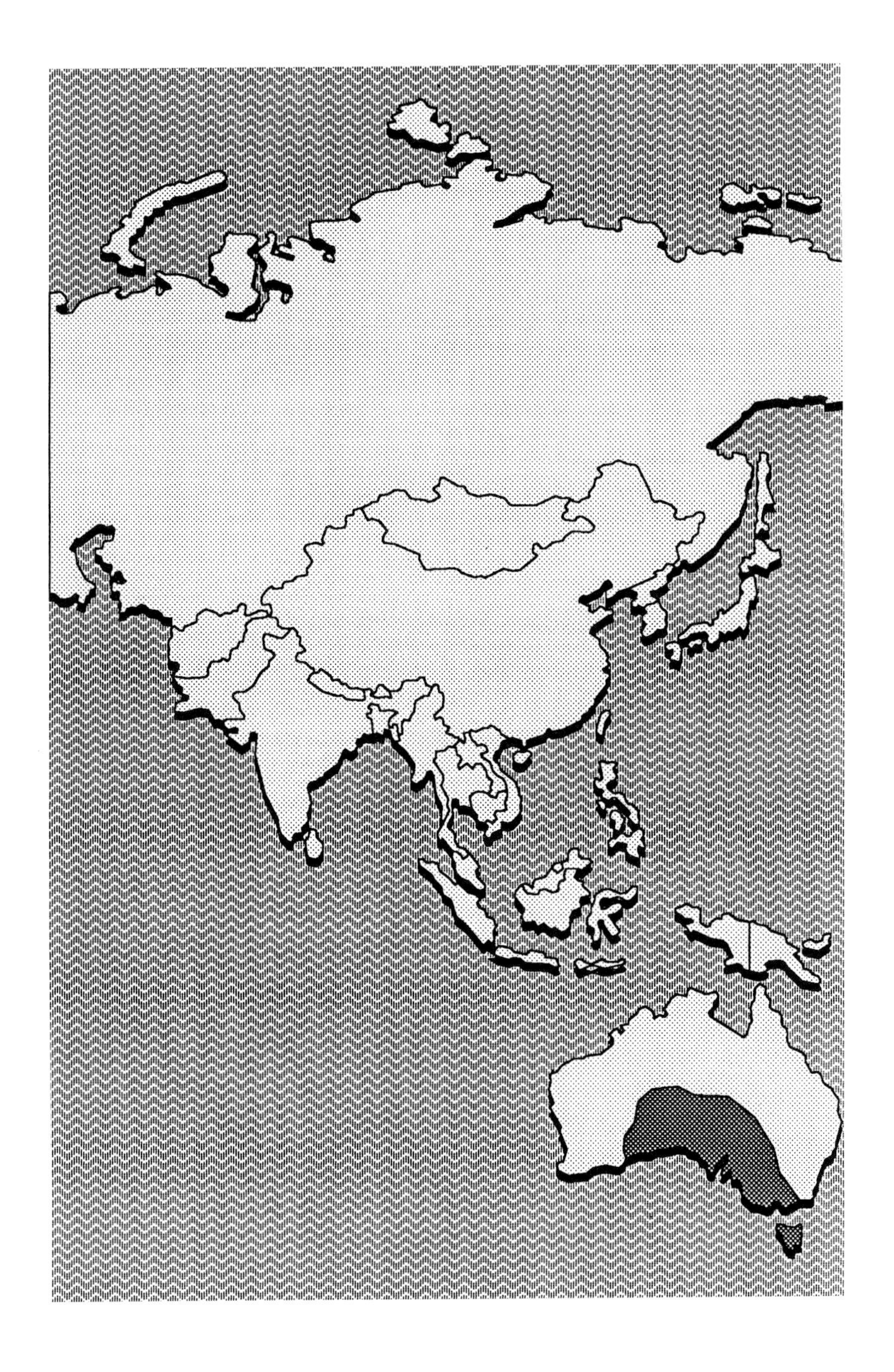

Death adder

Acanthopis antarcticus

Description: Reddish, yellowish, or brown color with distinct dark brown crossbands. The end of its tail is black, ending in a hard spine.

Characteristics: When aroused, this highly dangerous snake will flatten its entire body, ready to strike over a short distance. It is nocturnal, hiding by day and coming out to feed at night. Although it has the appearance of a viper, it is related to the cobra family. Its venom is a powerful neurotoxin; it causes mortality on about 50 percent of the victims, even with treatment.

Habitat: Usually found in arid regions, fields, and wooded lands.

Length: Average 45 centimeters, maximum 90 centimeters.

Distribution: Australia, New Guinea, and Moluccas.

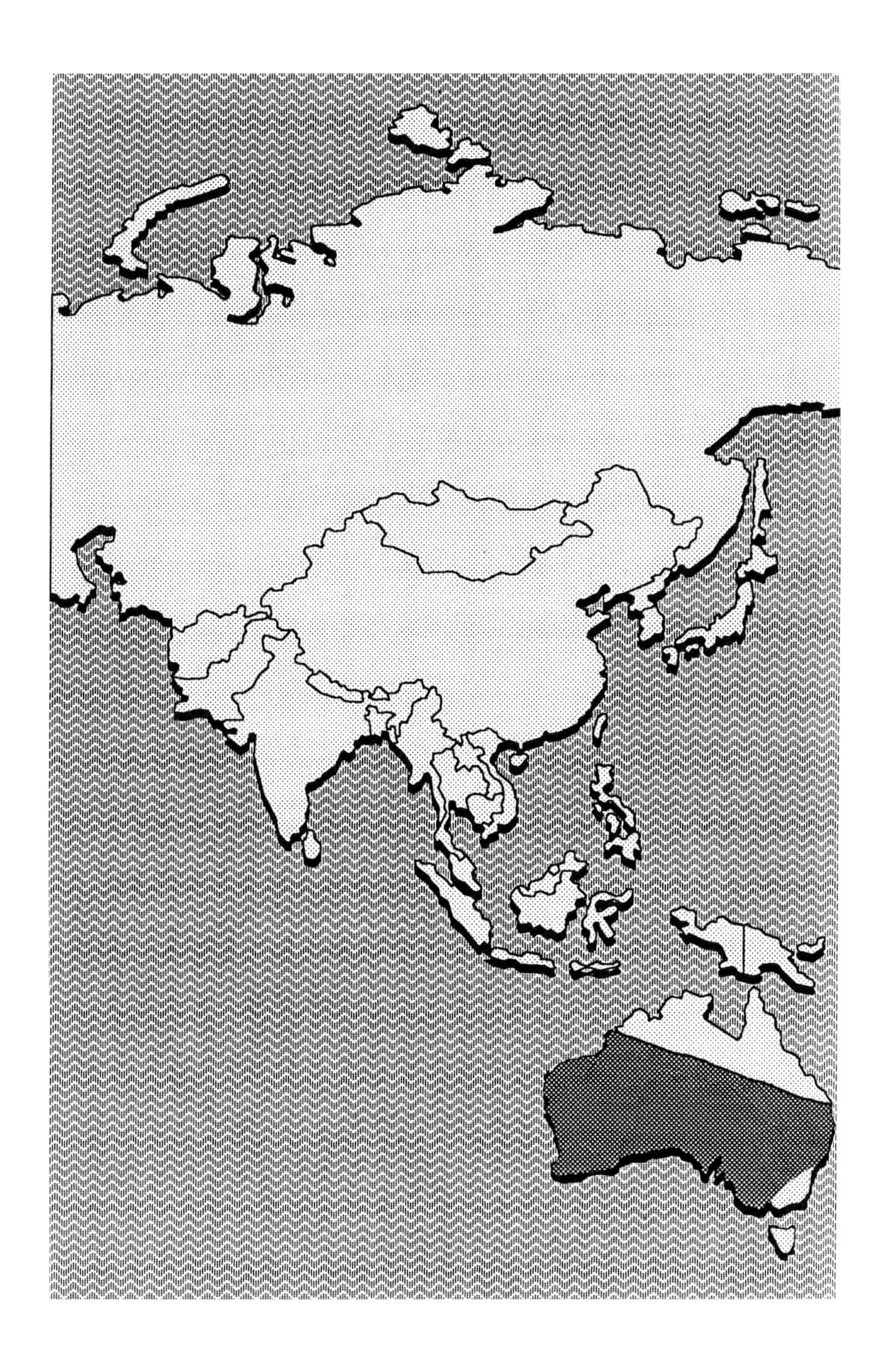

Taipan

Oxyuranus scutellatus

Description: Generally uniformly olive or dark brown, the head is somewhat darker brown.

Characteristics: Considered one of the most deadly snakes. It has an aggressive disposition. When aroused, it can display a fearsome appearance by flattening its head, raising it off the ground, waving it back and forth, and suddenly striking with such speed that the victim may receive several bites before it retreats. Its venom is a powerful neurotoxin, causing respiratory paralysis. Its victim has little chance for recovery without prompt medical aid.

Habitat: At home in a variety of habitats, it is found from the savanna forests to the inland plains.

Length: Average 1.8 meters, maximum 3.7 meters.

Distribution: Northern Australia and southern New Guinea.

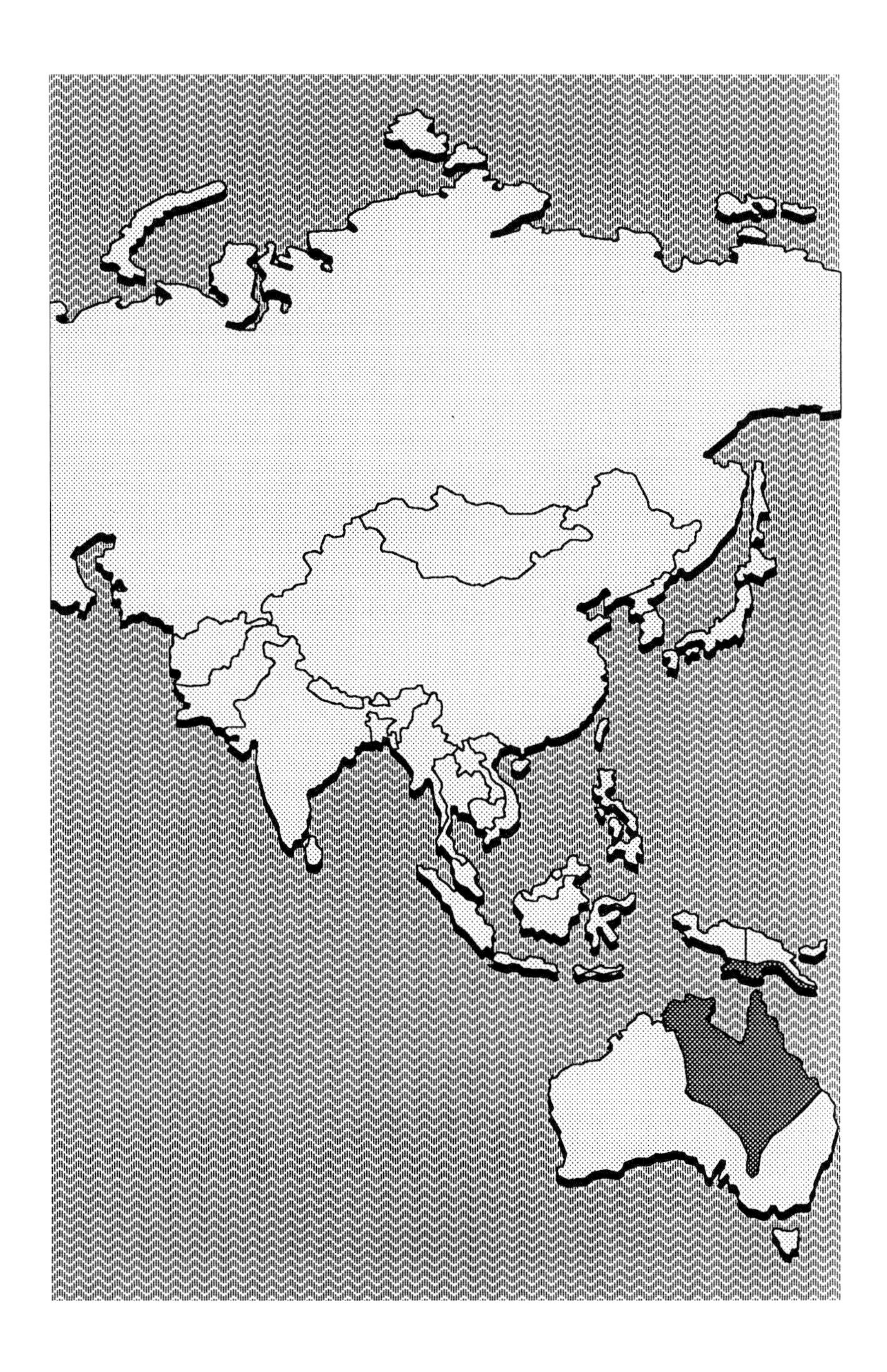

Tiger snake

Notechis scutatus

Description: Olive to dark brown above with yellowish or olive belly and crossbands. The subspecies in Tasmania and Victoria is uniformly black.

Characteristics: It is the most dangerous snake in Australia. It is very common and bites many humans. It has a very potent neurotoxic venom that attacks the nervous system. When aroused, it is aggressive and attacks any intruder. It flattens its neck making a narrow band.

Habitat: Found in many habitats from arid regions to human settlements along waterways to grasslands.

Length: Average 1.2 meters, maximum 1.8 meters.

Distribution: Australia, Tasmania, Bass Strait islands, and New Guinea.

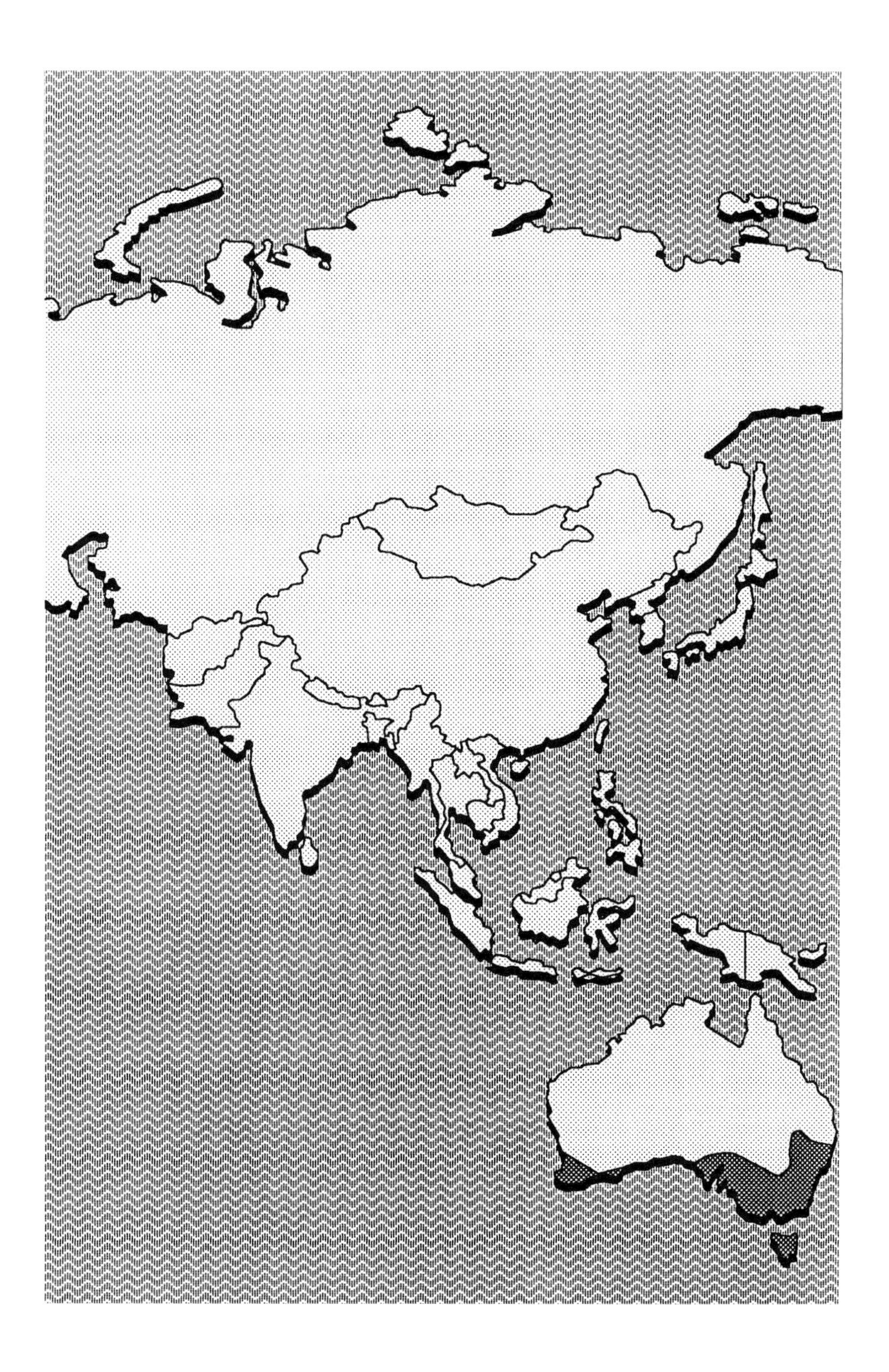

POISONOUS SEA SNAKES

Banded sea snake
Laticauda colubrina

Description: Smooth-scaled snake that is a pale shade of blue with black bands. Its oarlike tail provides propulsion in swimming.

Characteristics: Most active at night, swimming close to shore and at times entering tide pools. Its venom is a very strong neurotoxin. Its victims are usually fishermen who untangle these deadly snakes from large fish nets.

Habitat: Common in all oceans, absent in the Atlantic Ocean.

Length: Average 75 centimeters, maximum 1.2 meters.

Distribution: Coastal waters of New Guinea, Pacific islands, the Philippines, Southeast Asia, Sri Lanka, and Japan.

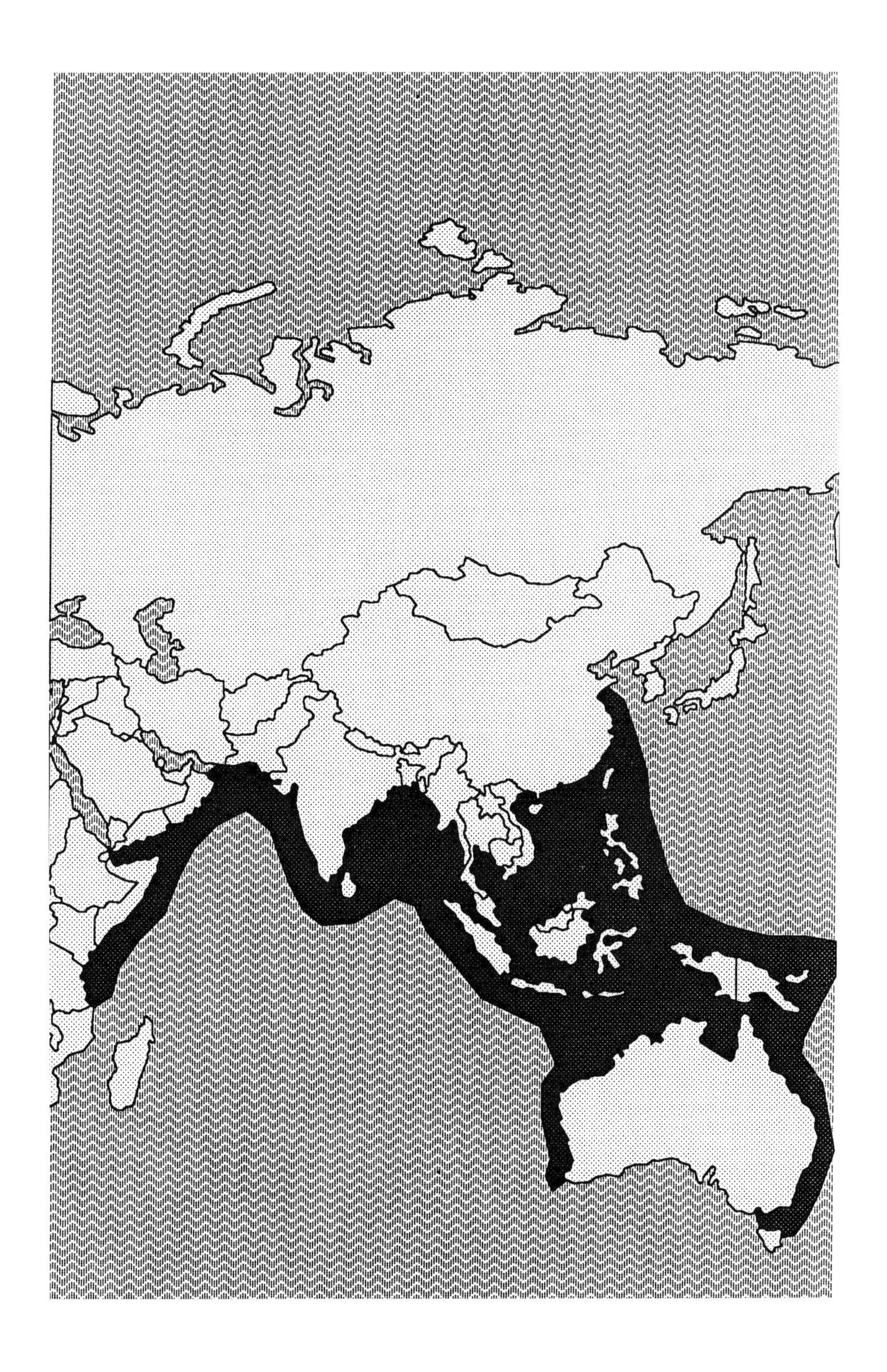

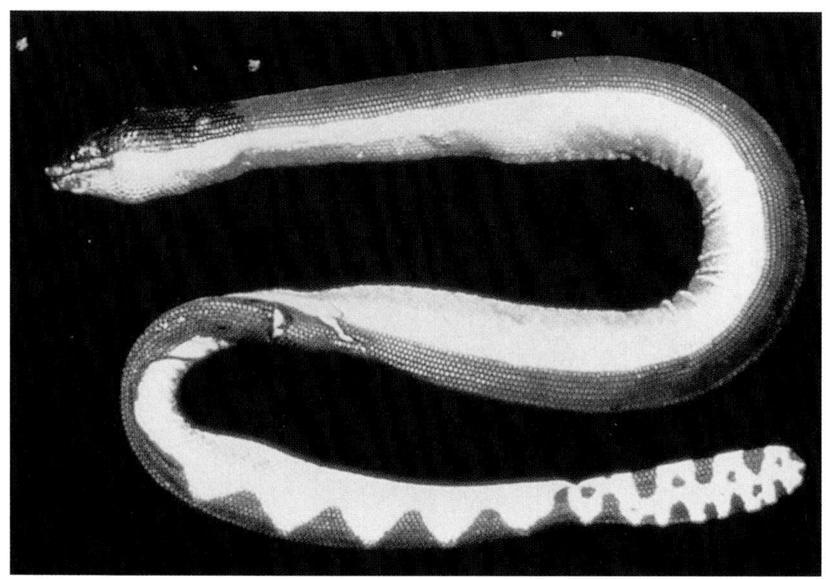

WAIKIKI AQUARIUM

Yellow-bellied sea snake
Pelamis platurus

Description: Upper part of body is black or dark brown and lower part is bright yellow.

Characteristics: A highly venomous snake belonging to the cobra family. This snake is truly of the pelagic species – it never leaves the water to come to shore. It has an oar-like tail to aid its swimming. This species is quick to defend itself. Sea snakes do not really strike, but deliberately turn and bite if molested. A small amount of their neurotoxic venom can cause death.

Habitat: Found in all oceans except the Atlantic Ocean.

Length: Average 0.7 meter, maximum 1.1 meters.

Distribution: Throughout the Pacific Ocean from many of the Pacific islands to Hawaii and to the coast of Costa Rica and Panama.

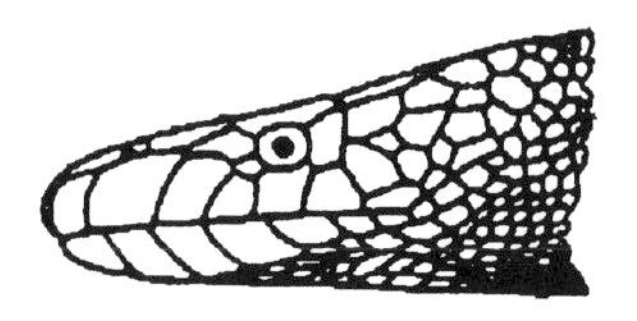

POISONOUS LIZARDS

Gila monster

Heloderma suspectum

Description: Robust, with a large head and a heavy tail. Its body is covered with beadlike scales. It is capable of storing fat against lean times when food is scarce. Its color is striking in rich blacks laced with yellow or pinkish scales.

Characteristics: Not an aggressive lizard, but it is ready to defend itself when provoked. If approached too closely, it will turn toward the intruder with its mouth open. If it bites, it hangs on tenaciously and must be pried off. Its venom glands and grooved teeth are on its bottom jaw.

Habitat: Found in arid areas, coming out at night or early morning hours in search of small rodents and bird eggs. During the heat of the day it stays under brush or rocks.

Length: Average 30 centimeters, maximum 50 centimeters.

Distribution: Arizona, New Mexico, Utah, Nevada, northern Mexico, and extreme corner of southeast California.

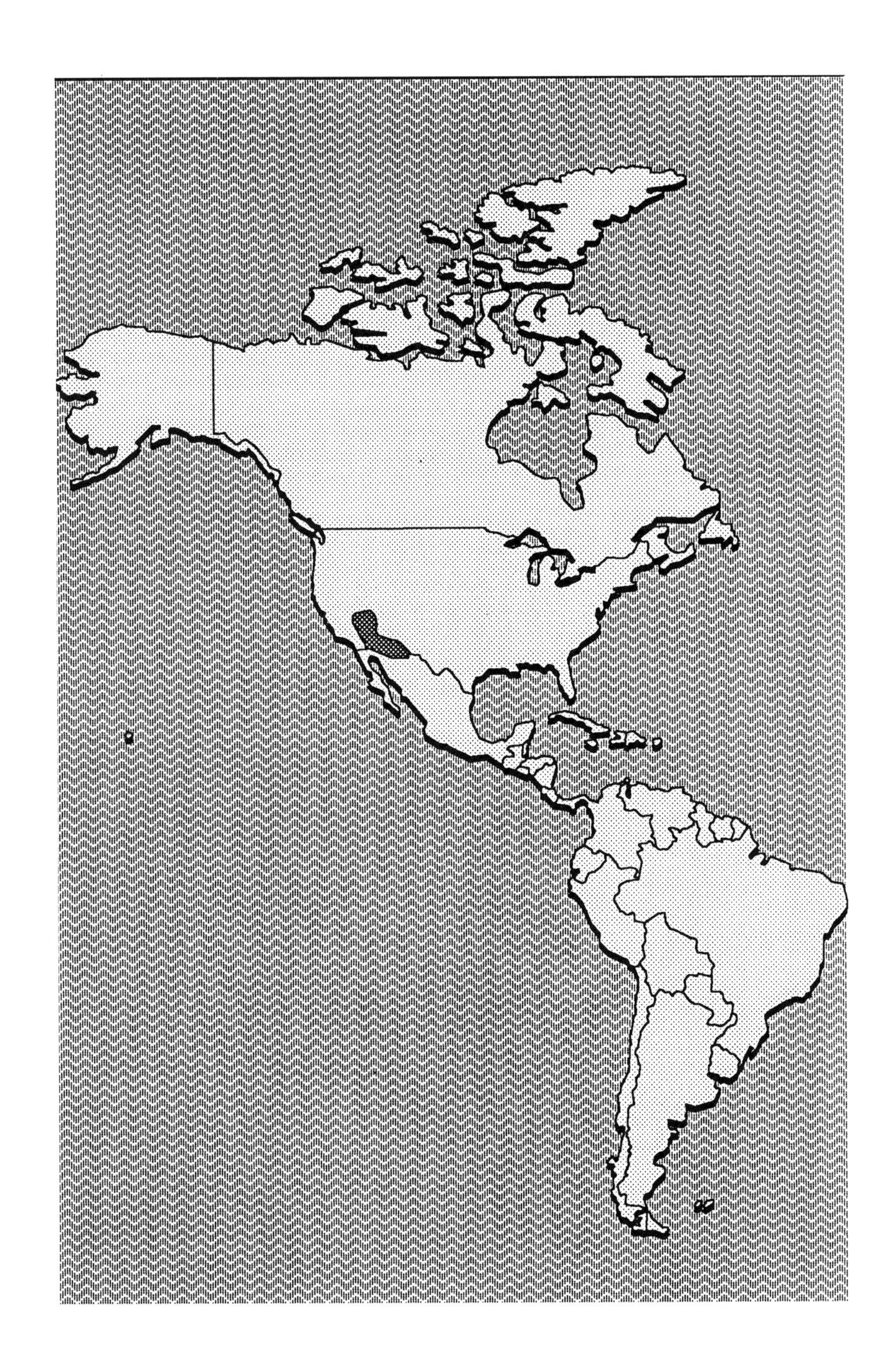

JOHN H. TASHJIAN/FORT WORTH ZOO

Mexican beaded lizard
Heloderma horridum

Description: Less colorful than its cousin, the gila monster. It has black or pale yellow bands or is entirely black.

Characteristics: Very strong legs let this lizard crawl over rocks and dig burrows. It is short-tempered. It will turn and open its mouth in a threatening manner when molested. its venom is hemotoxic and potentially dangerous to man.

Habitat: Found in arid or desert areas, often in rocky hillsides, coming out during evening and early morning hours.

Length: Average 60 centimeters, maximum 90 centimeters.

Distribution: Mexico through Central America.

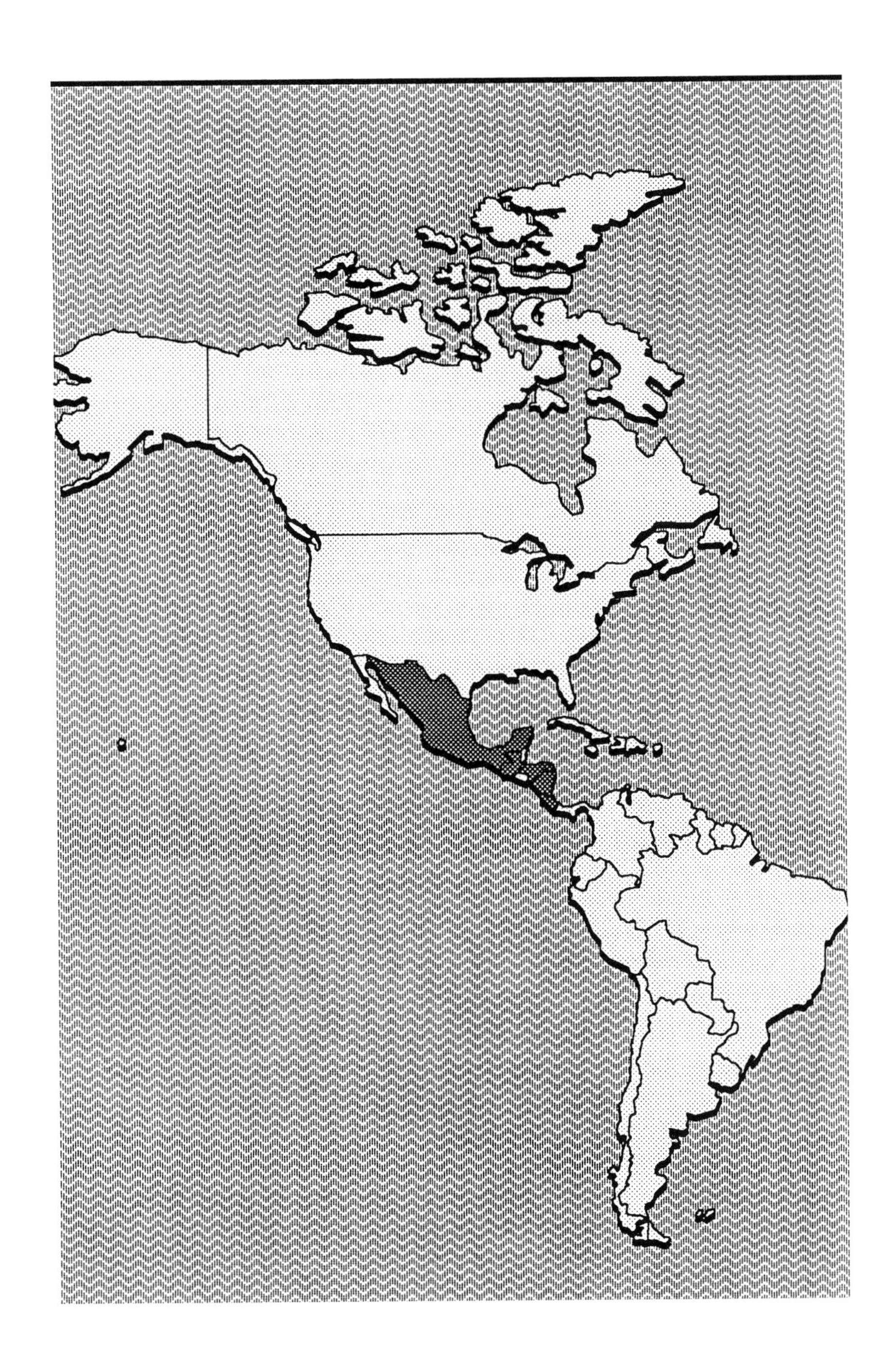

INDEX

D

E

T

U

V

W

Y